Elements of
Exterior Ballistics

Elements of Exterior Ballistics
LONG RANGE SHOOTING

FIRST EDITION

GEORGE KLIMI

Copyright © 2016 by George Klimi.

Library of Congress Control Number: 2016902024
ISBN: Hardcover   978-1-5144-5767-2
        Softcover   978-1-5144-5766-5
        eBook   978-1-5144-5765-8

All rights reserved. No part of this book may be reproduced or transmitted in any form or by any means, electronic or mechanical, including photocopying, recording, or by any information storage and retrieval system, without permission in writing from the copyright owner.

Any people depicted in stock imagery provided by Thinkstock are models, and such images are being used for illustrative purposes only.
Certain stock imagery © Thinkstock.

Print information available on the last page.

Rev. date: 03/07/2016

To order additional copies of this book, contact:
Xlibris
1-888-795-4274
www.Xlibris.com
Orders@Xlibris.com
730788

Contents

Author ix
Acknowledgment xi
Preface xiii
Short Description xv

Chapter 1
A Brief Approach to Aiming with Small Arms 1
    Introduction 1
    1.1 Projectile Ballistic Trajectory 1
    1.2 Ballistic Trajectory, Bullet Drop and Departure Angle 6
    1.3 Trajectory Height above the Horizontal Line 13
    1.4 Aiming Angle, Angle of Sight-Scope and Departure Angle 16
    1.5. Change of Zero Range 17
    1.6 Testing and Adjusting Aiming Angle 28
    1.7 Coordinate System Associated with Line of Sight 35
    1.8 Height of Trajectory over Line of Sight 39
    1.9 Change of Zero Range Using Bullet Path 42
    1.10 Change of Zero Range Using Trajectory Height 45
    1.11 Trajectory Height and Bullet Path Tables 48

Chapter 2
Long Range Inclined Shooting 54
    Introduction 54
    2.1 A Practical Method on Inclined Shooting 55
    2.2 Basic Formula for Inclined Shooting 61
    2.3 Drop, Super Elevation Angle and Inclined Range 64
    2.4 Rigidity Principle of Projectile Trajectory 67
    2.5 Non-Rigidity Principle of Trajectory 69
    2.6 Shooting with Departure Angle Zero 72
    2.7. Numerical and Approximate Solutions 74
    2.8 Exterior Ballistics PC Programs 78

Chapter 3
Standard Atmosphere in Exterior Ballistics 90
    Introduction 90
    3.1 Standard Atmosphere 91
    3.2 Characteristics of Atmospheric Air 92
    3.3 Change of Air Characteristics with Altitude 99
    3.4 Barometric Formula 103

Chapter 4
Elementary Exterior Ballistics 107
   Introduction 107
   4.1 Modified Piton-Bressant Trajectory 108
   4.2 Exponential Equation of Projectile Trajectory 114
   4.3 Exponential Equation in Non-Standard Atmosphere 123
   4.4 Modified Exponential Equation of Trajectory........................... 126
   4.5 Effect of Cartridge Temperature on Projectile Trajectory 131
   4.6 Approximate Equations to Estimate the Coriolis Effect 137
   4.7 Determination of Coriolis Effect, Spin Drift and Wind Effect 142
   4.8 Range-Wind and Cross-Wind 145
   4.9 Newton-Snell's Law in Exterior Ballistics 148
   4.10 Tangent Law on Similar Trajectories 153

Chapter 5
Differential Equations of Exterior Ballistics 158
   Introduction 158
   5.1 Differential Equations of Projectile Trajectory 159
   5.2 Reference G-Functions of Resistance 162
   5.3 Characteristic G-functions of Resistance 164
   5.4 Characteristic G-function Versus Reference G-function 168
   5.5 Estimation of Coriolis Effect 175
   5.6 Lift Force and Overturning Moment on Spinning Projectile 178
   5.7 Gyroscopic Stability Factor and Twist rate 181
   5.8 Estimation of Projectile Spin Deflection 186
   5.9 Improved Euler's Method Applied in Exterior Ballistics 190
   5.10 Measurement of Muzzle Velocity 200

Appendix A—Reference G-functions of Resistance 205
Appendix B—Characteristic G-functions 211
Appendix C—PC Programs 214
Appendix D—PC Programs 221
Appendix E—PC Programs 245
Appendix F—Algorithm of Improved Euler's Method Trajectory 264
Appendix G—Excel program on IEM 272
Appendix H—Electronic Copies of PC Program 283

Disclaimer 285
References 287
End Notes 291

To My Special Son Iven

To My Son Erio and to My Wife Dorina

AUTHOR

George Klimi (Gjergj Klimi), Ph.D., is a retired associate professor of Mathematics who taught Math at NYC College of Technology and at Pace University (New York).

George is the author of four books on exterior ballistics: "Exterior Ballistics with Applications" (2008), "Exterior Ballistics of Small Arms" (2009), Exterior Ballistics: A New Approach (2010), Exterior Ballistics: The Remarkable Methods (2014)

The third edition of "Exterior Ballistics with Applications", Xlibris, is published in December 2011.

George Klimi, from 1970 until 1993, has been professor of Physics & Mathematics and chair of Physics Department at the Military Academy of Tirana, Albania.

In 90's George used to work at the Committee of Science and Technology, and at the Ministry of Higher Education and Research in Albania as Director of TEMPUS Office, responsible for the implementation of the European Community Programs to restructure the higher education.

He was awarded the Gold Medal of Eagle, by the President of Republic of Albania, for his contribution in the democratic movement against dictatorship, and for restructuring the higher education in Albania.

ACKNOWLEDGMENT

My heartfelt gratitude is owed to my friend and colleague Col. Genc Kokoshi, ex-professor at the Academy of General Staff, Tirana, for the ingenious work in designing the PC programs related with my exterior ballistics books.

There are three main Exterior Ballistics PC programs associated with the book. The PC programs (codes), in quick basic (QB), were compiled in 1991 by Col. Genc Kokoshi. They are modified, time after time by the author, to reflect the advancements in Exterior Ballistics.

The core of all PC programs that are developed since 2005 are the code's that Col. Kokoshi has rigorously designed in 1991.

The three initial PC programs were based on Siacci's G-function and were valid only for a type of artillery projectile, when field cannon shooting is done in TSA atmosphere.

The QB codes designed by Col. Kokoshi are fascinating compact codes that made possible for me to advance in the study of exterior ballistics.

PREFACE

"Elements of Exterior Ballistics: Long Range Shooting" is a concise but comprehensive instructive book on exterior ballistics applied into long range shooting with small arms.

The foundations of the book are innovatively related to the exterior ballistics of point-mass projectile as well as to the new findings and contemporary ballistics methods presented in my preceding books.

The book is designed for exterior ballistics professionals, amateurs, and competitive shooters interested in long range shooting and, in general, in exterior ballistics.

Though the exterior ballistics applications are related to long range shooting with small arms, the reader can easily extend the ballistics techniques to the artillery fire.

Physicists, mathematicians and engineers have a lot of interesting applied topics that are not present in traditional physics and engineering textbooks.

A good knowledge in Calculus and Physics is necessary to fully understand and make use of the special techniques the exterior ballistics uses to predict the projectile trajectory.

The readers with a good background in Pre-calculus have no difficulty to comprehend the topics exposed in the first four chapters of the book.

In addition, the large number of examples, in each topic, demonstrates the exterior ballistics solving techniques and helps the reader to understand the ballistics concepts and principles, as well as the challenging theoretical and practical applications.

The PC programs associated with the book are a great tool to the study of exterior ballistics, to predict the projectile trajectory and aim the firearm to the center of the target.

Each topic is methodically and rigorously presented to avoid ambiguity that is present in the contemporary exterior ballistics, especially in non-mathematical or non-technical presentation of exterior ballistics.

George Klimi, PhD
New York, January 2016
iven24@aol.com,

SHORT DESCRIPTION

"Elements of Exterior Ballistics: Long Range Shooting", (EEB), consists of five chapters.

The first two chapters deal exclusively with horizontal long range shooting as well as with mountain and inclined shooting.

Knowledge in algebra and trigonometry is necessary for the marksman to understand and apply the techniques of inclined shooting and of zeroing in the firearm at the target.

In chapter two, there is introduced the non-rigidity principle that is a beautiful, simple, and thrilling technique applied in practice of long range and inclined shootings.

We introduce the "Non-Rigidity-Principle" that can be used to find the super elevation or super depression angle and to set up the aiming angle in uphill or downhill shooting.

The non-rigidity principle was been used in quite all my books without mentioning it.

At the end of the second chapter are included many interesting exercises that are solved with the use of PC programs. The reader can use the PC programs also to verify the accuracy of different methods that we have used to predict the projectile trajectory.

The PC programs include as well the techniques to predict the effect of range-wind and cross-wind.

The prediction method of wind effects are based on the Didion's approach that uses the relative motion and relative velocity of a projectile in presence of range wind or cross-wind. In other words, to study the wind effect there is considered an observer that moves with the velocity of range-wind, or with velocity of cross-wind.

The third chapter provides a review of standard atmospheres and the techniques used to calculate the characteristics of atmosphere at shooting sites.

The forth chapter, "Elementary Exterior Ballistics", includes elementary equations that describe the projectile trajectory in standard and non standard atmosphere.

Though the elementary trajectory equations (third order parabola, modified exponential equation) are approximate, the prediction accuracy of the elements of the ballistics trajectory we obtain employing those equations is remarkable.

In this chapter are presented approximate techniques to estimate the Coriolis Effect and cross-wind drift, as well as some methods that allow the long range shooter to measure, in practice of shooting, the wind drift, the spin drift, and the Coriolis Effect.

The remarkable Newton-Snell's law and Tangent Law on similar trajectories (included in chapter four) are some simple techniques to predict accurately the elements of bullet trajectory in a non-standard atmosphere when we know the standard firing range tables.

Particularly, those two laws give the possibility to the shooter to easily calculate the departure angle, as well as the aiming angle to zero in the rifle in non standard atmosphere when we know the range table in standard atmosphere, and vice versa.

The fifth chapter contains the differential equations of point-mass trajectories that can be solved numerically when we know the ballistics coefficient related to reference G_1-function, G_7- functions or when it is known the characteristic G-function of an individual bullet obtained by Doppler radar measurements, or other methods.

The end of the chapter contains the study: Improved Euler's Method Applied in Exterior Ballistics.

It is a surprise to find out that the Improved Euler's method (IEM), known also as Heun's method, gives accurate results for relatively large steps of integration when we solve the differential equations of projectile trajectory.

This unexpected outcome is contradictory to the belief, claimed in differential equations textbooks, that IEM gives accurate integration results only for small integration steps.

1
A Brief Approach to Aiming with Small Arms

Introduction

This chapter contains the methods related with zeroing in a rifle at a given range.

There is shown mathematically the relationship between the coordinate system related with the horizontal range and the rectangular system of coordinates related with the line of sight (LOS).

There are presented the relations between projectile drop, departure angle and bullet path. Those relations are mathematically described by simple equations.

There are shown as well some methods we use to change the zeroing in.

1.1 Projectile Ballistic Trajectory

The traditional exterior ballistics employs simplified mathematical models to study the flight of projectiles by ignoring some factors that have insignificant influence in the accuracy of the practical results and temporarily neglecting some other factors (wind, projectile gyroscopic effect, Coriolis effect, etc.) whose influence is reflected later, in the obtained results of ballistic trajectory, as corrections.

The only forces that act on the projectile, considered by the simplified mathematical model, include the drag force and gravity. The result of the action of the drag force and the gravity on the projectile is the ballistic trajectory.

The ballistics trajectory is the flight course of an unguided point-mass projectile taken under the action of drag force and gravity after the projectile leaves the muzzle of firearm with a given initial velocity till it hits the target, or the ground.

The exterior ballistics studies the ballistic trajectory with respect to a three **dimensional system of Cartesian coordinates** xoyz that has the origin at the muzzle of the firearm, or at the sea level, while x-axis is along the horizontal line. The direction of x-axis is along the projection of the initial velocity $\vec{v}_0$ on the horizontal line.

In general, the muzzle of firearm is at a point with coordinates (0, y0), i.e. at the altitude y0 over the sea level.

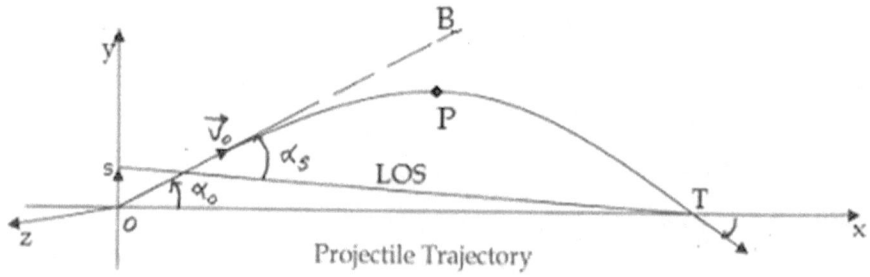

"Figure 1"

The ballistic trajectory of point-mass projectile is plane, i.e. the flying projectile does not deviate from the launching plane xoy (there is no crosswind or other lateral deviations due to the rotation of the projectile, etc.).

The ballistics trajectory is a set of points with coordinates (x, y, z = 0). The coordinates of the projectile are functions of time of flight (t), i.e.
$x = x(t)$, $y = y(t)$, $z = z(t) = 0$.

At any point, (x, y) that corresponds to the time of flight t the projectile has a defined velocity $\vec{v}$ tangent to the trajectory, i.e. a defined speed v and an angle α that the velocity forms with x-axis.

The projectile trajectory has the following characteristic elements:

The departure angle α_0, the initial velocity (muzzle speed) v_0, the ballistics coefficient c, the horizontal range $x_T = OT$ (or the coordinates of the point of impact ($x_T, y_T, z_T = 0$) to the target), the time of flight to the target t_T, the terminal speed v_T, the angle of impact on the target α_T and the maximum trajectory altitude y_m (or the coordinates of the trajectory vertex ($x_m, y_m, z_m = 0$)).

The target (T) can be at any point over, under, or at the altitude of the firearm.

Initial Velocity and Exit Velocity of Projectile

When a round is fired, the gases of the propellant charge exit the muzzle of the firearm at a velocity greater than the bullet exit velocity (for some bullets around 50% greater).

The hot and high pressure gases, expanding and interacting with the bullet, increase the projectile velocity beyond the exit velocity at the muzzle, until the bullet travels some centimeters to few meters (that depend on the type of projectile (bullet) propellant charge and firearm).

At the end of this transition period, the bullet has reached the maximum velocity (this velocity is not the initial velocity).

The **initial velocity** of a bullet (projectile) is a **fictive velocity** that makes the projectile follow the real trajectory if, at the muzzle, the projectile is launched at the **fictive velocity**, assuming that at the muzzle the powder gases cease to increase exit velocity of bullet beyond the muzzle.

For the Initial velocity there are in use terms: departure velocity, muzzle velocity, departure speed.

The muzzle velocity of a standard projectile, fired from a brand new barrel, depends on the propellant temperature.

For the ICAO atmosphere and for the ASM atmosphere the standard temperature of propellant charge is 21.11 degree Celsius (70 degree Fahrenheit.

The standard temperature of the propellant charge for the TSA atmosphere is 15 degree Celsius (59 degree Fahrenheit).

The propellant temperature, which is usually equal to the temperature of air at the shooting site, changes the initial standard velocity of the projectile when the temperature of air at the firing site is different from the temperature of the standard atmosphere.

For that reason, some standard range tables, which are valid for firing in a standard atmosphere with a standard firearm and projectile, contain the range corrections for a given change in propellant temperature.

Initial Velocity and Temperature of Cartridge

The actual initial velocity of a projectile v_0, for a given temperature T of propellant charge, can be estimated using the equation

$$v_{0T} = v_0 \cdot [1 + 0.001 \cdot (T - T_0)], \quad (1.1.1)$$

where v_0 is the muzzle standard velocity of the projectile that corresponds to the standard temperature T_0 of propellant (21.11 degree Celsius, for ICAO and ASM atmospheres; 15 degree for TSA atmosphere).

Note that for different types of black powder, instead of coefficient 0.001 we might have other values (see Exterior Ballistics: The Remarkable Methods", page p. 453)

Initial Velocity and Bullet Mass

For the same cartridge and propellant charge, any change in bullet mass, however small, changes the exit velocity of the bullet as well as the initial velocity of bullet, according to the equation (see Exterior Ballistics with Application, 3rd ed, page 457, Xlibris 2011)

$$v_{0m} = v_0 (1 - 0.4 \frac{dm}{m_0}). \quad (1.1.2)$$

where m_0 is the standard mass of bullet, $dm = m - m_0$ is the change in bullet mass.

Thus, if the standard launching velocity of a bullet is 850m/s, a relative increase of $dm/m_0 = 0.01 = 1\%$ in bullet mass changes the initial velocity from 850m/s to

$$v_{0m} = v_0 (1 - 0.4 \frac{dm}{m_0}) = 850 \cdot (1 - 0.4 \cdot 0.01) = 846 m/s.$$

Note that both formulas (1.1.1) and (1.1.2) are considered in the PC programs associated with the book.

Initial Velocity and Barrel Length

The muzzle velocity of a given bullet changes with the change of barrel length.

The change in muzzle velocity Δv_0 with a change ΔL_0 in barrel length can be estimated approximately by the equation (See Exterior ballistics: The Remarkable Method, page 455)

$$\Delta v_0 = 0.28916 \cdot v_0 \cdot \Delta L_0. \tag{1.1.3}$$

Hence, for the velocity that corresponds to the increased or decreased length we have

$$v_{0B} = v_0 [1 + 0.28916 \cdot (L - L_0)]. \tag{1.1.4}$$

Thus, if the departure velocity of a bullet is 830m/s then for an increase of $\Delta L_0 = L - L_0 = 0.05m$ in barrel length the muzzle velocity will be

$$v_{0B} = v_0 [1 + 0.28916 \cdot (L - L_0)] = 830 \cdot (1 + 0.28916 \cdot (0.05)) = 842 m/s.$$

Example 1.1 The standard initial velocity of a projectile fired in ICAO atmosphere is 860 m/s (if we consider that the projectiles are stored in a temperature 21.11 degree Celsius).

In shooting situations, it is usually not possible to keep the temperature of propellant charge equal to 21.11 degree Celsius.

Assume that the actual temperature of air at the firing site is 10 degree Celsius.

What will be the departure velocity of the projectile if the temperature of projectile cartridge (propellant charge) is the same as the actual temperature of air?

Solution

Using (1.1.1), we find that the departure velocity of the projectile is
$$V_0 = v_0 \cdot (1 + 0.001 \cdot (T - T_0)) = 860 \cdot (1 + 0.001 \cdot (10 - 21.11)) = 850.54 \ m/s.$$

The change in initial velocity is
$$\Delta(v_0) = V_0 - v_0 = 850.54 - 860 = -9.46 \ m/s.$$

1.2 Ballistic Trajectory, Bullet Drop and Departure Angle

Geometric Elements of Ballistic Trajectory

The geometric elements of the ballistic trajectory of a projectile, with respect to a rectangular Cartesian system of coordinates are (see fig. 1 and fig 2):

- The horizontal line (ox) is the line that originates at the muzzle of the firearm and has the direction of the horizontal component of initial velocity.
- Line of Departure (LOD) is the tangent line to the ballistic trajectory at the muzzle of the firearm and has the direction of the initial velocity ($\vec{v}_o$) of the projectile. The initial speed of the projectile is v_o. It is evident that the departure line is the direction of the axis of the muzzle of firearm just as the projectile leaves the muzzle.
- Angle of departure (α_0) is the angle measured from x-axis to the line of departure.
- Line of sight (LOS) is the straight line that connects the muzzle of the firearm and the target (center of the target or a point on the target. For small arms or sight firearms, the LOS is the line that connects the eye of the marksman, the scope, and the target. LOS is neither parallel to the axis of the bore, nor to the horizontal line (x-axis).
  Note that LOS changes the direction when we change the range of the target.
- Aiming angle is the angle measured from LOS to the departure line LOD=OB.
- In inclined shooting, the elevation angle is the angle measured from the x-axis to the line that connects the origin of coordinates (muzzle of firearm) and the target located on the inclined plane.
- Super elevation angle is the angle that zeroes the rifle on the inclined range of the target located on the inclined plane.
- The **horizontal range** $x_T = OT$ is the distance from the muzzle of the firearm to the target T, when both the target and the muzzle of firearm are at the same altitude over the sea level ($y_0 = y_T = 0$).
- Projectile drop at a given point P on the trajectory is the vertical distance QP from the line of departure to the point P (fig. 2).

Bullet Drop

The drop of a projectile $\overline{y}_D = QP$ (fig. 2), at a given time and location P, during the flight, is the perpendicular distance of bullet measured from the line of departure OB to the point P on the trajectory (fig. 2).

The projectile drop in the coordinate system xoy is a negative number, i.e. $\overline{y}_D < 0$.

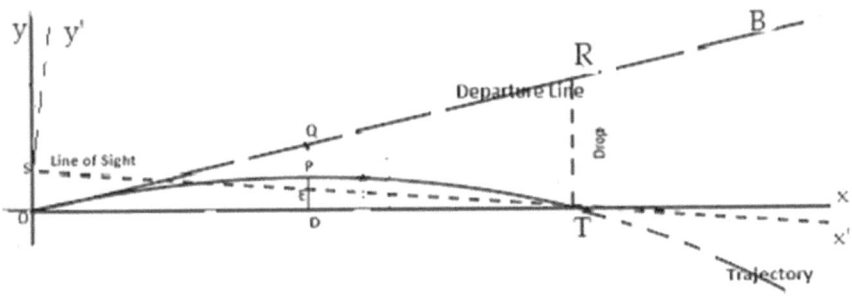

"Figure 2"

Departure Angle

In some standard range tables it is given the bullet drop $\overline{y}_T$ at zero-range $x_T = OT$. Using the bullet drop $\overline{y}_T$ we can find the departure angle α_{0T} that zeroes in the firearm at $x_T = OT$.

The projectile drop that corresponds to the horizontal range $x_T = OT$ is $\overline{y}_T = RT$. The departure angle $\alpha_{OT} = \angle TOB$ (fig. 1, fig. 2) is the departure angle necessary to hit the target located at the horizontal range $x_T = OT$. In other words, α_{OT} is the departure angle that zeroes the firearm at $x_T = OT$.

From fig. 2, for the departure angle α_{OT}, we can write:

$$\tan \alpha_{0T} = \frac{|\overline{y}_T|}{x_T} = -\frac{\overline{y}_T}{x_T}. \tag{1.2.1}$$

Hence,

$$\alpha_{OT} = \arctan(\frac{|\overline{y}_T|}{x_T}). \tag{1.2.2}$$

Note that the bullet drop is a negative number, $\overline{y}_T < 0$.

Since in long range shooting the departure angles are usually very small, using (1.2.1), we can write approximately:

$$\alpha_{OT} \approx \frac{|\bar{y}_T|}{x_T} \cdot \frac{180}{\pi}, 1 \qquad (1.2.3)$$

where α_{OT} is measured in degree.

For the departure angle in MOA (Minutes of Angle), using equation (1.2.3) we can write:

$$\alpha_{OT} \approx \frac{|\bar{y}_T|}{x_T} \cdot \frac{10,800}{\pi}. \qquad (1.2.4)$$

Bullet Drop and Departure Angle

Using (1.2.1), for the projectile drop (in absolute value), at the horizontal range $x_T = OT$ we have:

$$|\bar{y}_T| = x_T \cdot \tan \alpha_{OT} \qquad (1.2.5)$$

or, since the drop is negative,

$$\bar{y}_T = -x_T \cdot \tan \alpha_{OT}. \qquad (1.2.6)$$

To find the drop $\bar{y}_D = QP$ of the projectile at any point P on the trajectory, when the projectile is lunched at the angle α_{OT} that zeroes the firearm at $x_T = OT$, we need to know the height $y_D = DP$ of the trajectory at the point P with abscissa x_D. The projectile drop, $\bar{y}_D = QP$, at the point P with abscissa x_D (fg. 2) is

$$\bar{y}_D = y_D - y_Q = y_D - x_D \tan \alpha_{OT}, \qquad (1.2.7)$$

where y_D is the y-coordinate of the projectile located at a point P on the trajectory.

The quantity y_D is at the same time the height of bullet over the horizontal line ox (when bullet is launched with angle α_{OT} that zeroes the firearm at $x_T = OT$).

Dividing both sides of (1.2.7) by x_D, we have:

$$\frac{\bar{y}_D}{x_D} = \frac{y_D}{x_D} - \tan \alpha_{OT}. \qquad (1.2.8)$$

Note that the range $x_D = OD$ can be smaller or greater than the zero-range $x_T = OT$.

The left side of (1.2.8) is the tangent of the departure angle α_{0D} that zeroes the rifle at the horizontal range $x_D = OD$

Thus, for the departure angle α_{0D} that zeroes in the rifle at $x_D = OD$, we have:

$$-\tan \alpha_{0D} = \frac{y_D}{x_D} - \tan \alpha_{0T}. \qquad (1.2.9)$$

Hence,

$$\tan \alpha_{0D} = \tan \alpha_{0T} - \frac{y_D}{x_D}, \qquad (1.2.10)$$

or approximately:

$$\alpha_{0D} \approx \alpha_{0T} - \frac{y_D}{x_D} \cdot \frac{180}{\pi}, \qquad (1.2.11)$$

Hence, for the change in departure angle, when there is a change in zero range, we have:

$$\alpha_{0D} - \alpha_{0T} \approx -\frac{y_D}{x_D} \cdot \frac{180}{\pi}. \qquad (1.2.12)$$

Denoting

$$\Delta \alpha_{OT} = (\alpha_{0D} - \alpha_{0T}), \qquad (1.2.13)$$

the change in respective ranges, we can write (1.2.12):

$$\Delta \alpha_{OT} \approx -\frac{y_D}{x_D} \cdot \frac{180}{\pi}. \qquad (1.2.14)$$

NOTES

- **Radian and Degree Measures**

  In international system of units (SI), the angles are measured in radian.

  In the book we measure the angles mostly in degree.

To convert an angle measured in radian into degree, we multiply the angle in radian by $(180/\pi)$ and, vice versa, by $(\pi/180)$.

- **MOA**

  Another unit used in long range shooting is a minute of angle (MOA).

  To convert an angle measured in degree into MOA we multiply the angle in degree by 60.

  To convert an angle measured in radian into an angle in degree we use the multiplication factor $(180/\pi)$.

  To convert an angle measured in radian into an angle in MOA we use the multiplication factor $10,800/\pi$.

- **Approximate Trigonometric Equations**

  Throughout the book, since the departure angles in long range shooting with small arms are relatively small, close to zero degree, we use the approximate equations:

(a) For angle α measured in radian

$$\sin\alpha \approx \alpha, \quad \tan\alpha \approx \alpha, \qquad (1.2.15)$$

(b) For angle α measured in degree

$$\sin\alpha \approx \alpha \cdot \frac{\pi}{180}, \quad \tan\alpha \approx \alpha \cdot \frac{\pi}{180}. \qquad (1.2.16)$$

Thus, for example, the solution of the equation

$$\sin\alpha = 0.0262$$

is

$$\alpha = \arcsin(0.0262) = 1.50132°.$$

Employing (1.2.15), for the approximate solution we can write:

$$\alpha \cdot \frac{\pi}{180} \approx \sin\alpha = 0.0262.$$

Hence,

$$\alpha \approx 0.0262 \cdot \frac{180}{\pi} = 1.50115°.$$

Example 2.1 Departure Angle

The drop of 0.338 Lapua GB528 Scenar 19.44g bullet at the horizontal range $x_T = 1,500m$ is $\bar{y}_T = -30.035m$.

Find the departure angle needed to zero the firearm at the given range.

Solution

Employing (1.2.1), we find that

$$\alpha_{OT} = -\arctan(\frac{\bar{y}_T}{x_T}) = -\arctan(\frac{-30.035}{1500}) = 1.1471°.$$

Employing (1.2.3), we get quite the same departure angle:

$$\alpha_{OT} = -\frac{\bar{y}_T}{x_T} \cdot \frac{180}{\pi} = -(\frac{-30.035}{1500}) \cdot \frac{180}{\pi} = 1.1473°.$$

Example 2.2 Projectile Drop

The rifle, firing a 0.338 Lapua GB528 Scenar 19.44g bullet, is zeroed in at the horizontal range, $x_T = 1,200$ meters. The departure velocity of bullet is 830m/s (2723.10 fps).

According to table 3.1 section 5.3, the departure angle that zeroes the rifle at $x_T = 1,200$ meters is $\alpha_{0T} = 0.7838°$.

Find the projectile drop at horizontal range x = 1,200 meters.

Solution

Substituting in (1.2.3) we find that the bullet drop is

$$\bar{y}_T = -x_T \cdot \tan \alpha_{0T} = -1200 \cdot \tan(0.7838) = -16.42m.$$

Example 2.3 Projectile Drop

The rifle firing a 0.338 Lapua GB528 Scenar 19.44g bullet is sighted in at the horizontal range $x_T = 1,200$ meters. The departure angle that zeroes the rifle at $x_T = 1,200$ meters is $\alpha_{0T} = 0.7838°$ (see table 3.1 section 5.3). The departure velocity of bullet is 830m/s.

Find the projectile drop at the range $x_D = 900$ meters, if the height of the projectile over the horizontal line is $y_D = 4.24m$.

Solution

Using (1.2.7), we find that the projectile drop is

$$\bar{y}_D = y_D - x_D \cdot \tan \alpha_{0T} = 4.24 - 900 \cdot \tan(0.7838) = -8.072m.$$

Note

The drop, estimated above, is the same as the drop calculated when the rifle is zeroed at 900 meters.

Indeed, the departure angle that zeroes the rifle at $x_D = 900$ meters is $\alpha_D = 0.5126°$ (see table 3.1, section 5.3). Employing (1.2.6) we find the drop:

$$\bar{y}_D = -x_D \cdot \tan \alpha_D = -900 \cdot \tan(0.5126) = -8.052m.$$

Example 2.4 Predicting Departure Angle

A rifle fires a 0.338 Lapua GB528 Scenar 19.44g bullet, with departure velocity 2723 fps in ICAO atmosphere. The departure angle that zeroes the rifle at $x_T = 1,000$ yards, is $\alpha_{0T} = 0.5247°$.

At $x_D = 800$ yards, the height of the bullet trajectory above horizontal line is $y_D = 68.70$ inches.

Find the departure angle α_{0D} that zeroes the firearm at $x_D = 800$ yards.

Solution

Substituting in (1.2.11), we find that the departure angle that zeroes the rifle at $x_D = 800$ yards is

$$\alpha_{0D} \approx \alpha_{0T} - \frac{y_D}{x_D} \cdot \frac{180}{\pi} = 0.5247 - \frac{68.70}{800 \cdot (36)} \cdot \frac{180}{\pi} = 0.3380°.$$

Example 2.5

The rifle firing a 0.338 Lapua GB528 Scenar 19.44g bullet with velocity 830 m/s, is zeroed at the horizontal range $x_{OT} = 600$ meters.

At $x_D = 300$ meters the height of the trajectory above the horizontal line is $y_D = 0.872m$.

What is the change in departure angle if we change the zero range from $x_{OT} = 600$ to $x_D = 300$ meters?

Find the change in departure angle.

Solution

Using (1.2.12) we find that the change in departure angle is

$$(\alpha_{0D} - \alpha_{0T}) \approx -\frac{y_D}{x_D} \cdot \frac{180}{\pi} = -\frac{0.872}{500} \cdot \frac{180}{\pi} = -0.0999°.$$

1.3 Trajectory Height above the Horizontal Line

The trajectory height $y_D = DP$ above the horizontal line (ox) usually is given in standard range tables. It can be obtained solving the system of differential equations (2.7.1), or using other methods.

Let's estimate the projectile height over (below) the horizontal line ox. We assume that the rifle is zeroed in at range $x_T = OT$ and α_{0T} is the departure angle corresponding to the zero-range $x_T = OT$.

If the projectile height $y_D = DP$, at horizontal range $x_D = OD$ is not known, then using (1.2.7), for the y-coordinate of the projectile height at the given range x_D we have:

$$y_D = x_D \tan \alpha_{0T} + \bar{y}_D, \qquad (1.3.1)$$

where drop $\bar{y}_D < 0$. The equation (1.3.1) can be written:

$$y_D = x_D \cdot \frac{y_T}{x_T} + \bar{y}_D. \qquad (1.3.2)$$

Let's consider that we know the angles $\alpha_{0T} = \angle TOA$ and $\alpha_{0D} = \angle DOP$ that zero the firearm respectively at $x_T = OT$ and $x_D = OD$ (fig. 3 below)

The projectile height y_D at the range $x_D = OD$, when the firearm is zeroed at the range $x_T = OT$, is (fig. 2):

$$y_D = x_D \tan \alpha_0 - x_D \tan \alpha_D = x_D (\tan \alpha_{0T} - \tan \alpha_{0D}). \quad (1.3.3)$$

Using equation (1.3.3), we can write the equivalent equation:

$$y_D = x_D (\frac{|\bar{y}_T|}{x_T} - \frac{|\bar{y}_D|}{x_D}). \quad (1.3.4)$$

Equation (1.3.3) can be written in approximate form as:

$$y_D \approx x_D (\alpha_{0T} - \alpha_{0D}) \cdot \frac{\pi}{180}. \quad (1.3.5)$$

Note that the point D can be in front of point T, or behind T respectively when range $x_D = OD$ is smaller or greater than zero-range $x_T = OT$.

Example 3.1 Trajectory height over the horizontal line

The rifle firing a 0.338 Lapua GB528 Scenar 19.44g bullet is zeroed at the horizontal range $x_{OT} = 1,000$ meters. The corresponding departure angle that zeroes the rifle at 1,000 meters is $\alpha_{0T} = 0.5958°$ (see table 3.1, section 5.3).

(a) Find the height of the bullet trajectory at $x_D = 900$ meters.
(b) Find the bullet height at range $x_D = 1,100$ meters.

Solution

(a) In table 3.1 section 5.3, we find that the departure angle that zeroes the firearm at $x_D = 900$ meters is $\alpha_{0D} = 0.5126°$. Employing equation (1.3.3), we find that the bullet (at 900 meters) passes

$$y_D = x_D (\tan \alpha_{0T} - \tan \alpha_{0D}) = 900 \cdot (\tan 0.5958 - \tan 0.5126)$$
$$= 1.31m$$

over the horizontal line (fig. 3).

(b) In table 3.1, section 3.5, we find that the departure angle that zeroes the firearm at $x_D = 1,100$ meters is $\alpha_{OD} = 0.6858°$. Employing equation (1.3.3) we have:

$$y_D = x_D (\tan \alpha_{OT} - \tan \alpha_{OD}) = 1100 \cdot (\tan 0.5958 - \tan 0.6858)$$
$$= -1.728 m.$$

The y-coordinate (height) of bullet trajectory at $x_D = 1,100$ meters, when the firearm is zeroed at $x_{OT} = 1,000$ meters is negative, i.e. the bullet trajectory is under the horizontal line.

We get the same value using (1.3.5). Indeed,

$$y_D \approx x_D (\alpha_{OT} - \alpha_{OD}) \cdot \frac{\pi}{180} = 1100 \cdot (0.5958 - 0.6858) \cdot \frac{\pi}{180}$$
$$= -1.728.$$

Example 3.2 Trajectory height

A shooter with a Russian SKS rifle fires a 7.62mm bullet with departure speed 735m/s.

At what point over the horizontal line the bullet will hit a rectangular table-target located at a distance of $x_D = 100$ meters from the SKS in order that the firearm must be zeroed at $x_T = 300$ meters?

From the standard range table of the Russian rifle we find that the drop of the bullet at $x_D = 100$ meters and $x_T = 300$ meters is respectively $\bar{y}_D = -0.15m$ and $\bar{y}_T = -1.13m$.

Solution

Aiming at $x_T = 300$ meters, with the corresponding scope sight the gunman fires some bullets.

In accordance with equation (1.3.4), we expect that, at $x_T = 300$ meters, the vertical deviation of center of distributions of bullets from the bottom of the rectangular board-target to be

$$y_D = x_D (\frac{|\bar{y}_T|}{x_T} - \frac{|\bar{y}_D|}{x_D}) = 100 \cdot (\frac{1.13}{300} - \frac{0.15}{100}) = 0.227 m.$$

1.4 Aiming Angle, Angle of Sight-Scope and Departure Angle

Aiming Angle α_S is the angle measured from the LOS=ST to the departure line OB.

Angle of Sight-Scope is the angle $\alpha_h = \angle OTS$ measured from the horizontal line OT to LOS (fig, 1, fig. 2 above). **Angle of sight-scope is negative** since it is obtained by a clock-wise rotation.

Relationship between Aiming Angle, Departure Angle and Angle of Sight-Scope

Assume that the rifle is zeroed in at range $x_T = OT$.

If the scope height of the rifle is $y_S = h_T$, and we aim at $x_T = OT$ then the **angle of sight-scope**, $\alpha_h = \angle OTS$, is given by the equation:

$$\tan \alpha_{hT} = -(h_T / x_T), \qquad (1.4.1)$$

or, approximately by

$$\alpha_{hT} \approx -\frac{h_T}{x_T} \cdot \frac{180}{\pi}. \qquad (1.4.2)$$

The departure angle $\alpha_{0T} = \angle TOR$ zeroes the rifle at range $x_T = OT$. The aiming angle α_{ST} is given by the formula:

$$\alpha_{ST} = \alpha_{0T} - \alpha_{hT}, \qquad (1.4.3)$$

which can be written in compact form:

$$\alpha_S = (\frac{|\bar{y}_T|}{x_T} + \frac{h_T}{x_T}) \cdot \frac{180}{\pi}. \qquad (1.4.4)$$

Example 4.1

A Sierra NATO bullet caliber 0.30, 168 grain HPBT, fired horizontally, at the sea level, with initial speed 2,650 fps at the horizontal range $x_T = 600$ yards.

At $x_T = 600$ yards, the bullet drop of is $\bar{y}_T = 126.50$ inches.

(a) Find the angle of departure needed to zero the gun at a horizontal range of $x_T = 600$ yards from the gun.

(b) Find as well the aiming angle that the line of bullet departure forms with the line of sight (LOS). The sight-scope height is 1.5 inches.

Solution

(a) The departure angle α_{0T} that zeros the firearm at the horizontal range $x_T = 600$ yards is

$$\alpha_{0T} = \frac{|\bar{y}_T|}{x_T} \cdot \frac{180}{\pi} = \frac{126.50}{600 \cdot (36)} \cdot \frac{180}{\pi} = 0.3356° = 20.133 MOA.$$

(b) For the angle of sight (aiming angle) that the line of sight (LOS) forms with the horizontal line at the horizontal range 548.60m we can write:

$$\alpha_{hT} = -\frac{h_T}{x_T} \cdot \frac{180}{\pi} = -\frac{1.5}{600 \cdot (36)} \cdot \frac{180}{\pi} = -0.00398° = 0.2387 MOA.$$

The angle of sight is

$$\alpha_{ST} = \alpha_{0T} - \alpha_{hT} = 0.3356 - (-0.00398) = 0.3396° = 20.37 MOA.$$

Compact Solution

Substituting in (1.4.4), for the aiming angle we have:

$$\alpha_S \approx \frac{|\bar{y}_T| + h_T}{x_T} \cdot \frac{180}{\pi} = \frac{126.50 + 1.5}{600 \cdot (36)} \cdot \frac{180}{\pi} = 0.3395°.$$

1.5. Change of Zero Range

For small changes $\Delta\alpha_{0T}$ and $\Delta\alpha_{hT}$ respectively in departure angle α_{0T} and in angle of sight-scope α_{hT}, the aiming angle α_{ST} (defined in formula (1.4.3)) changes with the quantity $\Delta\alpha_S$ given by equation:

$$\Delta\alpha_{ST} = \Delta\alpha_{0T} - \Delta\alpha_{hT}, \tag{1.5.1}$$

where $\Delta\alpha_{hT} < 0$,

$$\Delta\alpha_{0T} = \alpha_{OD} - \alpha_{0T}, \tag{1.5.2}$$

and
$$\Delta\alpha_{hT} = \alpha_{hD} - \alpha_{hT}. \tag{1.5.3}$$

Thus, the change in aiming angle is
$$\Delta\alpha_{ST} = (\alpha_{OD} - \alpha_{OT}) - (\alpha_{hD} - \alpha_{hT}). \tag{1.5.4}$$

Compact Equivalent Equation

Let's express the above equations through the knowing drops and the angles of sight scopes.

Assume that the rifle is zeroed at a given range x_T and the corresponding departure angle is
$$\alpha_{0T} = \frac{|\bar{y}_T|}{x_T} \cdot \frac{180}{\pi}, \tag{1.5.5}$$

The angle of sight-scope is
$$\alpha_{hT} = -\frac{h_T}{x_T} \cdot \frac{180}{\pi}. \tag{1.5.6}$$

Aiming angle at x_T is
$$\alpha_{ST} \approx (\frac{|\bar{y}_T|}{x_T} + \frac{h_T}{x_T}) \cdot \frac{180}{\pi}. \tag{1.5.7}$$

Note that h_T is the sight-scope height when the zero range is x_T.

Without changing the sight-scope height h_T, we change the LOS by aiming at the point D that is at the range $x_D = OD$.

At the same time we change the departure angle from α_{OT} to α_{OD}, where α_{OD} correspond to range $x_D = OD$.

Though the scope height does not change, the angle of sight-scope changes from the value α_{hT} given in (1.5.6) to the value
$$\alpha_h = -\frac{h_T}{x_D} \cdot \frac{180}{\pi}. \tag{1.5.8}$$

Thus, the new aiming angle is

$$\alpha_{SDT} \approx (\frac{|\bar{y}_D|}{x_D} + \frac{h_T}{x_D}) \cdot \frac{180}{\pi}. \qquad (1.5.9)$$

The actual change in aiming angle is

$$\Delta\alpha_{SDT} \approx [(\frac{|\bar{y}_D|}{x_D} - \frac{|\bar{y}_T|}{x_T}) + (\frac{h_T}{x_D} - \frac{h_T}{x_T})] \cdot \frac{180}{\pi}. \qquad (1.5.10)$$

Now, we keep unchanged the first term of (1.5.10) and adjust the height of sight-scope at the new zero-range $x_D = OD$. The height of sight-scope at new zero range is denoted h_D.

As result of two successive operations, the total change in aiming angle is

$$\Delta\alpha_{ST} \approx [(\frac{|\bar{y}_D|}{x_D} - \frac{|\bar{y}_T|}{x_T}) + (\frac{h_D}{x_D} - \frac{h_T}{x_T})] \cdot \frac{180}{\pi}. \qquad (1.5.11)$$

where

$$\alpha_{hD} = -\frac{h_D}{x_D} \cdot \frac{180}{\pi}. \qquad (1.5.12)$$

Thus (1.5.11) estimates the correction in aiming angle needed to zero the firearm at the zero-range $x_D = OD$.

The change in aiming angle, given in (1.5.11), in practice is performed by changing the sight-scope angle with the quantity

$$\Delta\alpha_h = \Delta\alpha_{ST}. \qquad (1.5.13)$$

Thus, the change in sight-scope angle is

$$\Delta\alpha_{hT} \approx [(\frac{|\bar{y}_D|}{x_D} - \frac{|\bar{y}_T|}{x_T}) + (\frac{h_D}{x_D} - \frac{h_T}{x_T})] \cdot \frac{180}{\pi}. \qquad (1.5.14)$$

The equation (1.5.14) is equivalent to equation (1.5.4).

Another Compact Formula

Assuming that the rifle is zeroed in at $x_T = OT$. We want to change the zeroing in at $x_D = OD$ when we know the bullet trajectory height $y_D = DP$ (at range x_D).

Employing (1.2.12), (1.2.14), (1.5.4) and (1.5.14), we find that the change in sight-scope angle can be calculated using the following equation:

$$\Delta\alpha_{hT} \approx (-\frac{y_D}{x_D} + \frac{h_D}{x_D} - \frac{h_T}{x_T}) \cdot \frac{180}{\pi}. \qquad (1.5.15)$$

The angle of sight-scope that zeroes the rifle at $x_D = OD$, is

$$\alpha_{hD} = \alpha_{hT} + \Delta\alpha_{hT} \qquad (1.5.16)$$

where α_{hT} and $\Delta\alpha_{hT}$ are estimated using respectively (1.5.6) and (1.5.14 or (1.5.15).

Height of Sight-Scope

The height of sight-scope h_D that zeroes the firearm at $x_D = OD$ is calculated employing (1.5.6), and (1.5.14), i.e. solving the equation:

$$\alpha_{hT} + \Delta\alpha_{hT} = -\frac{h_D}{x_D} \cdot \frac{180}{\pi}. \qquad (1.5.17)$$

Number of Clicks to Adjust the Scope-Sight Angle

Usually the sight adjustment is done when zeroing the rifle at a certain horizontal range.

Every telescopic sight scope shows the number of clicks the marksman has to dial to change the sight-scope angle by a quantity in MOA.

For an iron sight, the rate of change in sight clicks depends on the distance between the front sight and rear sight.

For example, telescope sights usually are set up to change the sight-scope angle at a

$$rate = (1/4)\ MOA = (0.25) MOA \text{ (for 1 click)}.$$

Thus the number of clicks, corresponding to the change (1.5.14) or (1.5.15) is:

$$N\_clicks = \frac{(\Delta\alpha_{hT})}{rate}. \qquad (1.5.18)$$

Example 5.2 illustrates the correction of the angle of sight-scope.

Example 5.1 Aiming angle

A marksman fires a 0.30 ball M2 bullet with an initial speed of 2800 ft/s.

The drop of the bullet at the horizontal range $x_D = 100$ yards and $x_T = 500$ yards from the gun is respectively $\bar{y}_D = -2.80$ inches and $\bar{y}_D = -76.50$ inches.

(a) Where does the bullet hit a target board located at a distance of $x_D = 100$ yards from the gun if the firearm is zeroed at $x_T = 500$ yards?

(b) Show as well, the aiming angle that zeroes in the rifle at $x_T = 500$ yards.

The sight-scope height that zeroes the rifle at 500 m is 1.5 inches.

Solution

(a) Substituting in (1.3.4), we find that the trajectory height at $x_D = 100$ yards is

$$y_D = x_D \left(\frac{|\bar{y}_T|}{x_T} - \frac{|\bar{y}_D|}{x_D} \right) = 100 \cdot \left(\frac{76.50}{500} - \frac{2.80}{100} \right) = 12.50 \ inches.$$

The point P that is $y_D = 12.50$ inches above the horizontal line is called **control point. Control Point P** controls the accuracy of zeroing the rifle at $x_T = 500$ yards, by measuring (at 100 yards) the vertical deviation of the group of shots from P.

(b) Substituting in (1.5.7) we find that the aiming angle that zeroes the rifle at $x_T = 500$ yards, and let the bullet pass $y_D = 12.50$ inches over the center of target located at $x_D = 100$ yards is

$$\alpha_{ST} \approx \left(\frac{|\bar{y}_T|}{x_T} + \frac{h_S}{x_T} \right) \cdot \frac{180}{\pi} = \left(\frac{76.5}{500 \cdot (36)} + \frac{1.5}{500 \cdot (36)} \right) \cdot \frac{180}{\pi} = 0.2483°.$$

The aiming angle in MOA is

$$\alpha_S = 0.2483 \cdot (60) = 14.898 MOA.$$

Example 5.2 Step by Step Aiming Adjustment

A rifle firing a 0.308 Sierra 165 grain Spitzer boattail bullet is zeroed at the range $x_T = OT = 250$ yards (scope height $h_T = 1.8$ inches). Departure line is OR (fig. 2).

According to Sierra[1], at zero range $x_T = 250$ yards the bullet drop is $\bar{y}_T = -16.61$ inches, while at zero-range $x_D = 100$ yards the bullet drop is $\bar{y}_D = -2.37$ inches.

(a) Find the departure angle and the aiming angle that zeros in the rifle at range $x_{OT} = 250$.

(b) Calculate the change in aiming angle, when we change the zero-range from $x_{OT} = 250$ yards to $x_D = 100$ yards.

Rate 1 click = 1/4 MOA (rate =0.25 MOA).

Solution
(a) The departure angle $\alpha_{OT} = \angle TOR$ needed to zero the firearm at $x_{OT} = 250$ yards is

$$\alpha_{OT} = \frac{|\bar{y}_T|}{x_T} \cdot \frac{180}{\pi} = \frac{16.61}{(36) \cdot 250} \cdot \frac{180}{\pi} = 0.1057°.$$

The angle of sight-scope at $x_T = 250$ yards is

$$\alpha_{hT} = -\frac{h_T}{x_T} \cdot \frac{180}{\pi} = -\frac{1.80}{250 \cdot (36)} \cdot \frac{180}{\pi} = -0.0115°.$$

Aiming angle is

$$\alpha_{ST} = \alpha_{OT} - \alpha_{hT} = 0.1057 - (-0.0115) = 0.1172°.$$

At $x_D = 100$ yards, with this scope set up, the bullet will pass above the horizontal line.

Substituting in (1.3.4), we find that the "height" $y_D = DP$ of bullet "above" the horizontal line at $x_D = 100$ yards is

$$y_D = x_D \left(\frac{|\bar{y}_T|}{x_T} - \frac{|\bar{y}_D|}{x_D} \right) = 100 \cdot \left(\frac{16.61}{250} - \frac{2.37}{100} \right) = 4.274 in.$$

So, if the rifle is zeroed at $x_T = 250$ yards then the bullet at $x_D = 100$ yards will pass $y_D = 4.274$ inches above the horizontal line.

(b) Assume that with the same scope height setup, $h_T = 1.8"$, we aim at the point D at the range $x_D = 100$ yards (LOS = DS, fig. 1).

The departure angle that zeroes the firearm at $x_D = 100$ yards is

$$\alpha_{OD} = \frac{|\bar{y}_{DT}|}{x_{OD}} \cdot \frac{180}{\pi} = \frac{2.37}{(36) \cdot 100} \cdot \frac{180}{\pi} = 0.0377°.$$

The corresponding sight-scope angle is

$$\alpha_{hD} = -\frac{h_T}{x_D} \cdot \frac{180}{\pi} = -\frac{1.80}{100 \cdot (36)} \cdot \frac{180}{\pi} = -0.02865°.$$

The aiming angle is

$$\alpha_{SD} = \alpha_{OD} - \alpha_{hD} = 0.0377 - (-0.02865) = 0.06635°.$$

As result, the change in aiming angle is

$$\Delta\alpha_S = \alpha_{SD} - \alpha_{ST} = -0.06635 - (-0.1172) = -0.05083°.$$

Consider equation (1.5.4),

$$\Delta\alpha_S = \Delta\alpha_{0T} - \Delta\alpha_{hT},$$

where $\Delta\alpha_S = -0.05083°$.

According to the above equation, for a fixed left side equal to $\Delta\alpha_S = 0.05085°$, we can change one or the other term on the right side.

We must keep constant the change in departure angle α_{OD} to hit the target at range $x_D = 100$ meters. To compensate we need to "increase" $\Delta\alpha_S$ to the value:

$$\Delta\alpha_S = -0.05083° = -3.050 MOA.$$

That means that to the negative angle $\alpha_{hT} = -0.0115°$ (clock-wise rotation) we must add the above negative change in angle of sight-scope.

The number of clicks, needed to correct the aiming angle, must be dialed in the direction that increases the height of sight scope.

$$N\_clicks = \frac{|\Delta\alpha_S|}{rate} = (\frac{3.050}{1/4}) = 12.20 \approx 12 \ clicks.$$

Employing (1.5.16) we find that the angle of sight-scope is

$$\alpha_{hD} = \alpha_{hT} + \Delta\alpha_{hT} =$$
$$= -0.0115° - 0.05085° = -0.06235° = -3.741 MOA.$$

The new angle is negative since it is obtained by a clock-wise rotation.

In absolute value it is bigger than the absolute value of the angle of sight-scope $\alpha_{hT} = -0.0115°$.

So, in order that departure angle remain unchanged, equal to $\alpha_{oD} = 0.0377°$, we have to increase the sight-scope angle by the quantity, in absolute value, $\Delta\alpha_h = |-3.050| = 3.050 MOA$.

The new sight scope height can be found using equation (1.5.17). Thus we can write:

$$-\frac{h_D}{x_D} \cdot \frac{180}{\pi} = -0.06235°.$$

Hence, the height of the adjusted sight scope is $h_D = 3.918$ inches.

Note

Sierra gives a change in angle of sight-scope of $\Delta\alpha_S = 3.05 MOA$, when the zero range changes from $x_T = 250$ yards to $x_D = 100$ yards.

Compact Solution Formula (1.5.14)

The change in angle of sight can be found using compact formula (1.5.14). Indeed, we have:

$$\Delta\alpha_h \approx [\frac{|\bar{y}_D|}{x_D} - \frac{|\bar{y}_T|}{x_T} + \frac{h}{x_D} - \frac{h}{x_T}] \cdot \frac{180}{\pi} =$$
$$= [\frac{2.37}{100 \cdot (36)} - \frac{16.61}{250 \cdot (36)} + \frac{1.8}{100 \cdot (36)} - \frac{1.8}{250 \cdot (36)}] \cdot \frac{180}{\pi} = -0.0508°.$$

Employing Compact Formula (1.5.15)

We obtain the same result using compact formula (1.5.15).

Indeed,

$$\Delta\alpha_h = (-\frac{y_D}{x_D} + \frac{h}{x_D} - \frac{h}{x_T}) \cdot \frac{180}{\pi} =$$

$$= [\frac{-4.274}{100 \cdot (36)} + \frac{1.8}{100 \cdot (36)} - \frac{1.8}{250 \cdot (36)}] \cdot \frac{180}{\pi} = -0.05083°.$$

Example 5.3 Number of clicks to change the zeroing in

A rifle firing a 0.308 Sierra 165 grain Spitzer boattail bullet is zeroed at the range $x_T = 100$ yards.

The height of sight-scope above the bore line is $y_S = 1.8$ *inches*.

Rate: 1 click = 1/4 MOA (rate =0.25 MOA).

(a) For the range $x_D = 250$ yards, find the "height" of the bullet trajectory "above" horizontal line if the bullet drop at $x_D = 250$ yards is $\overline{y}_D = -16.61$ inches.

At zero-range $x_T = 100$ yards the bullet drop is $\overline{y}_T = -2.37$ inches.

(b) Change the zeroing from zero range $x_T = 100$ yards to $x_D = 250$ yards, and estimate the number of clicks needed to adjust the angle of sight-scope.

(c) Calculate the sight-scope height when rifle is zeroed in at $x_D = 250$ yards.

Solution

(a) Rifle is zeroed at $x_T = 100$ yards.

Substituting in (1.3.4), we find that the "height" of bullet "over" the horizontal line at $x_D = 250$ yards is

$$y_D = x_D (\frac{|\overline{y}_T|}{x_T} - \frac{|\overline{y}_D|}{x_D}) = 250 \cdot (\frac{2.37}{100} - \frac{16.61}{250}) = -10.685 in.$$

(b) At $x_D = 250$ yards the trajectory of bullet is $y_D = -10.685$ inches under the horizontal line.

Employing (1.5.15), we find that the change in sight-scope angle is

$$\Delta\alpha_{hT} = (-\frac{y_D}{x_D} + \frac{h_t}{x_D} - \frac{h_T}{x_T}) \cdot \frac{180}{\pi} =$$

$$= [-\frac{-10.685}{250 \cdot (36)} + \frac{1.8}{250 \cdot (36)} - \frac{1.8}{100 \cdot (36)}] \cdot \frac{180}{\pi} = 0.05083°.$$

The same result we obtain employing the compact formula (1.5.14). Indeed:

$$\Delta\alpha_{hT} \approx [\frac{|\bar{y}_D|}{x_D} - \frac{|\bar{y}_T|}{x_T} + \frac{h_T}{x_D} - \frac{h_T}{x_T}] \cdot \frac{180}{\pi} =$$

$$= [\frac{16.61}{250 \cdot (36)} - \frac{2.37}{100 \cdot (36)} + \frac{1.8}{250 \cdot (36)} - \frac{1.8}{100 \cdot (36)}] \cdot \frac{180}{\pi} = 0.05083°.$$

So, the change in the angle of sight-scope, when we change the zeroing at $x_D = 250$ yards is

$$\Delta\alpha_{hT} = 0.05083° = 3.050 MOA.$$

The number of clicks, needed to correct the aiming angle,

$$N\_clicks = \frac{|\Delta\alpha_{hT}|}{rate} = (\frac{3.050}{1/4}) = 12.20 \approx 12 \ clicks$$

must be dialed in the direction that increases the sight scope height.

(c) The sight-scope angle at $x_T = 100$ yards was:

$$\alpha_{hT} = -\frac{h_T}{x_T} \cdot \frac{180}{\pi} = -\frac{1.8}{100 \cdot (36)} \cdot \frac{180}{\pi} = -0.02865°$$

The sight scope angle at $x_D = 250$ is

$$\alpha_{hD} = \alpha_{hT} + \Delta\alpha_{hT} =$$
$$= -0.02865° + 0.05083° = 0.02218°.$$

Substituting in equation (1.5.17):

$$\alpha_{hD} = -\frac{h_D}{x_D} \cdot \frac{180}{\pi},$$

we can write:

$$-\frac{h_D}{x_D} \cdot \frac{180}{\pi} = 0.02218°.$$

Substituting in above equation $x_D = [250 \cdot (36)]$ inches, and solving for h_D we find that the height of the sight-scope at $x_D = 250$ yards is $h_D = -3.484$ inches.

The sight-scope height $h_D = -3.484$ is a negative number because it corresponds to a positive sight-scope angle (counter clockwise rotation, see fig.2).

The absolute value, $|h_D| = 3.484$ inches, represents the height of the sight scope at zero range $x_D = 250$ yards.

Comment

When we change the zeroing in one direction (example 5.2) or in the opposite direction (example 5.3), the change in angle (in absolute value) of the sight-scope is the same, $\Delta\alpha_{hT} = 0.05083°$.

As we can see from example 5.2 and example 5.3, the sight-scope angle, respectively at $x_D = 100$ yards and $x_D = 250$ yards, is not the same ($\alpha_{hD} = -0.06623°$ and 3.918 inches against $\alpha_{hD} = 0.02218°$ and $h_D = 3.484$ inches).

The discrepancies are related to the fact that in range tables, for aiming angle calculations, there is given the height of trajectory or the trajectory path, based on an arbitrary chosen height of sight-scope; (see table 11.1 and table 11.2 section 1.11, the sight height $h_T = 0.04$ meters).

Thus, the marksman starts setting up his/her sight-scope at a given zero range, let's say 600 meters, using his own sight scope height that might be different from $h_T = 0.40$ meters.

1.6 Testing and Adjusting Aiming Angle

In practice of firing with small arms, it is necessary to test the sight-scope setup of the firearm to assure shooting accuracy.

Testing of the sight scope is done using firing tests. The firing tests can be performed, for example:

- After repairing or adjusting the iron sight scope, parts of a rifle, etc.
- To assure that the sight scope is set up correctly to zero the firearm at the desired range.
- When a competitive marksman fires in atmospheric conditions that are not standard. For example, in a new location that is different from the location the marksman usually fires.
- When, in general, the long range shooting is not performed in standard conditions (not in a standard atmosphere and the departure velocity of bullet might be different from the standard one, etc.).
- When the long range shooting data are obtained solving the system of differential equation using the G_1, or G_7 drag function and a fixed ballistic coefficient.

It is known that for long range shootings the accuracy of data: bullet drop, departure angle, velocity and time of flight, are not that accurate.

Inaccuracy of Bullet Drop that is Predicted Using reference G_1, or G_7 functions, and a Fixed BC.

Using reference G-functions, G_1, or G_7, to predict the drop of a bullet in long range shooting might result in relatively large errors.

Thus, for example, according to Wikipedia[2], at range 1,200 meters, the drop of 0.338 GB528 19.44g Lapua bullet (initial velocity 830m/s), predicted using G_1-drag function and BC = 0.785, is 16.073 meters.

The drop of the same bullet at the same range, predicted using Doppler radar measurements is 16.571.

So, theoretically there is a change of -0.50 meters (16.073 - 16.571) between the respective drops. It means that the bullet will impact in vertical direction 0.50 meters above the center of the target.

That relatively large vertical deviation needs to be corrected.

The following example illustrates the sight-scope testing and correction method.

Example 6.1 Testing procedure

Consider the standard atmosphere TSA (traditional Standard Atmosphere0.

A marksman with a Russian rifle Simonov SKS fires some 7.62mm bullets with a departure speed of 735m/s on a control board located $x_D = 100$ meters from the rifle.

Using the standard range table of SKS rifle the shooter finds that the departure angle that zeroes the rifle at $x_T = 300$ meters is $\alpha_{0T} = 0.216°$, while the departure angle that zeroes the rifle at $x_D = 100$ meters is $\alpha_{OD} = 0.084°$.

The rifle has a telescopic sight. The angle of sight-scope is set up to zero the rifle at the horizontal range $x_T = 300$ meters when the shooting conditions are standard.

The scope height $h_T = 0.0381$ meter corresponds to zero range $x_T = 300$ meters.

After shooting with some bullets, the center of the group of shots at $x_D = 100$ meters is $DQ = 0.250$ meters above the lower edge of a rectangular board (aiming point).

(a) How high "above" the **control point P** of the board target does the rifle shoot with the actual sight-scope set up at $x_T = 300$ meters?

(b) Is the sight scope of the rifle set up correctly to be zeroed at $x_T = 300$ meters?

(c) Adjust the sight height to zero the firearm at $x_T = 300$.

Solution

The y-coordinate, $y_D = DP$, of the center of the group of shots at $x_D = OD$ (at 100 meters) in standard firing conditions must be

$$y_D = x_D(\alpha_{0T} - \alpha_{OD}) \cdot \frac{\pi}{180} = 100 \cdot (0.216 - 0.084) \cdot \frac{\pi}{180} = 0.230m, \quad (1.6.1)$$

over the center of the target, and not $DQ = 0.250m$.

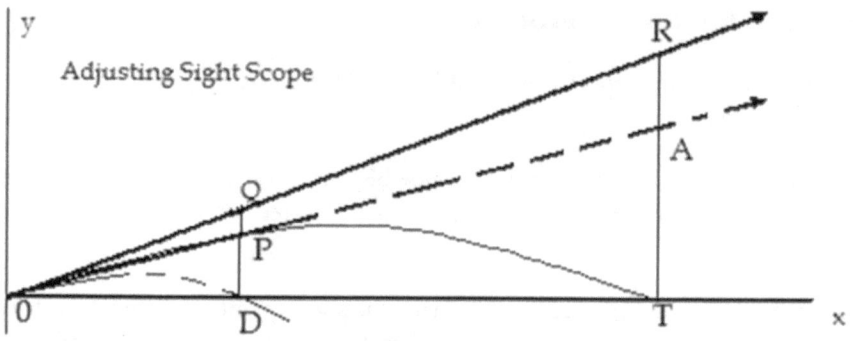

"Figure 3"

Control Point

The point P that is $y_D = 0.230$ meters above the horizontal line is called **control point**.

If the center of the group of shots, at range 100 meters, is at the point P then the sight scope is set up correctly, i.e. it zeroes the firearm at $x_T = 300$ meters.

The bullet will hit the center of the target at 250 meters.

Because of the causes listed above, the actual hit at the control board can be above or below the control point P.

(b) The rifle at 100 meters fires
$$PQ = DQ - DP = 0.250 - 0.230 = 0.02m$$
above the **control point P**.

That means that at $x_T = 300$ meters the rifles fires higher than it is predicted. The sight-scope is not set up correctly to hit the center of the target at $x_T = 300$.

The bullet, at $x_T = 300$, will hit above the center of the target.

(c) Adjusting Sight Scope Height

Correction of the sight-scope can be done by firing tests, using trial and error procedure, or by changing the height of sight employing theoretical procedure as it is shown below.

Since the angle of sight-scope, for the firearm zeroed at $x_T = 300$ is unchanged, the change in angle of sight-scope is $\Delta\alpha_h = 0$. Using (1.5.4), we find that the change in aiming angle is equal to the change in departure angle:

$$\Delta\alpha_S = \Delta\alpha_{0T}. \tag{1.6.2}$$

Figure 3 shows that the deviation of the departure angle from the range value given in standard range table is

$$\Delta\alpha_{0T} = (\frac{y_Q - y_D}{x_D}) \cdot \frac{180}{\pi}, \tag{1.6.3}$$

where $y_Q = DQ$, $y_D = DP$, P is the control point.

Substituting in (1.6.3) we have:

$$\Delta\alpha_{0T} = (\frac{y_Q - y_D}{x_D}) \cdot \frac{180}{\pi} = \frac{0.250 - 0.230}{100} \cdot \frac{180}{\pi} = 0.01146°$$

$$= 0.688 MOA.$$

Thus,

$$\Delta\alpha_S = \Delta\alpha_{0T} = 0.688 MOA. \tag{1.6.4}$$

To keep departure angle unchanged, i.e. in our example $\alpha_{0T} = 0.216°$, we have to reduce the angle of sight-scope by the correction quantity:

$$\Delta\alpha_h = \Delta\alpha_S$$

Hence, the correction (degree) in sight-scope is given by the compact formula:

$$\Delta\alpha_h = (\frac{y_Q - y_D}{x_D}) \cdot \frac{180}{\pi}. \tag{1.6.5}$$

Substituting in (1.6.5), we find the correction in angle of sight scope

$$\Delta\alpha_h = 0.688 MOA. \quad (1.6.6)$$

When $y_D = 0$, i.e. the control point is at the point with coordinates $(x_D, y_D = 0)$, using (1.6.3) we can write:

$$\Delta\alpha_D = (\frac{y_Q}{x_D}) \cdot \frac{180}{\pi}. \quad (1.6.7)$$

The equation (1.6.5) yields the change in angle of sight-scope,

$$\Delta\alpha_h = (\frac{y_Q}{x_D}) \cdot \frac{180}{\pi}. \quad (1.6.8)$$

when the control point is $(x_D, y_D = 0)$.

Applying Compact Equation to Correct Aiming Angle (sighting angle)

Assume that a marksman has set up the rifle scope to zero the firearm at x_D. To zero the firearm at x_D, the marksman has used the data obtained employing G_1-drag function and a fixed coefficient.

The marksman fires some bullets at the target located at range x_D.

The vertical deviation of the center of the group of shots, at x_D is PQ and not zero. The actual deviation of the departure angle from the departure angle that needed to zero the rifle at x_D is equal to the value estimated using (1.6.7):

$$\Delta\alpha_{0D} = \frac{QP}{x_D} \cdot \frac{180}{\pi} = (\frac{y_Q}{x_D}) \cdot \frac{180}{\pi}. \quad (1.6.9)$$

Thus, the change in sight-scope angle is estimated by the equation

$$\Delta\alpha_h = (\frac{y_Q}{x_D}) \cdot \frac{180}{\pi}. \quad (1.6.10)$$

Comment

The correction (1.6.5) of the sight-scope can be seen as result of the change in standard conditions of shooting (for example as result of changes in atmospheric conditions or other factors).

So, the aiming correction calculated in (1.6.5) is necessary to adjust the actual zero range that is greater than $x_T = 300$ meters.

Example 6.2 Shooting in non-standard conditions

A competitive shooter fires some 0.338 Lapua GB528 Scenar 19.44g bullet with a departure speed of 830 m/s on a control table located $x_D = 100$ meters from the rifle.

Using the standard range table 3.1 section 5.3, the shooter finds that the departure angle that zeroes the rifle at $x_T = 500$ meters is $\alpha_{0T} = 0.242°$, while the departure angle that zeroes the rifle at $x_D = 100$ meters is $\alpha_{0D} = 0.0419°$.

(a) Determine the control point P on the firing board, i.e. the height of the bullet trajectory above the horizontal line at $x_D = 100$ meters.

(b) Using ballistic tables, the shooter sets up the aiming angle in order to zero the rifle at $x_T = 500$ meters.

To test the scope set up, the shooter fires at a control table at $x_D = 100$ meters.

The center of the group of shoots is 0.360 meters above the horizontal line.

- Is the sight scope of the rifle set up correctly to be zeroed at $x_T = 500$ meters in the actual shooting site?
- How high is the bullet trajectory at $x_T = 500$ in the actual conditions?
- Adjust the sight height to zero the firearm at $x_T = 500$.

(a) Substituting in (1.3.4) we find the height of the bullet trajectory (at $x_D = 100$ meters):

$$y_D = x_D(\alpha_{0T} - \alpha_D) \cdot \frac{\pi}{180} = 100 \cdot (0.242 - 0.0419) \cdot \frac{\pi}{180}$$

$$= 0.349 m.$$

(b) The sight scope is not set up correctly.

The sight scope correction is

$$\Delta\alpha_h = \frac{y_Q - y_D}{x_D} \cdot \frac{180}{\pi} = (\frac{0.360 - 0.349}{100}) \cdot \frac{180}{\pi} = 0.00630°$$

$$= 0.378 MOA.$$

At $x_T = 500$ the bullet trajectory height is (fig. 3)

$$\Delta y_T = x_T \cdot \Delta \alpha_{0T} = x_T \cdot (\frac{y_Q - y_D}{x_D}) = 500 \cdot \frac{0.360 - 0.349}{100}$$

$$= 0.055m.$$

above the center of the target.

Example 6.3 Testing the Sight Setup

A marksman fires some 0.338 GB528 19.44g Lapua bullets (initial velocity 830m/s).

The marksman has zeroed in advance the rifle at $x_T = 1,200$ meters, using a set of trajectory heights (or trajectory paths) predicted by a PC program that uses the G_1 function of resistance with a ballistic coefficient BC = 0.785 (see section 1.11).

At the end of shooting, the marksman measures a vertical deviation of $y_Q = 0.50$ meters of the center of the group shots over the center of the target located at $x_T = 1,200$ meters.

At $x_T = 1,200$ meters, the PC predicted drop of projectile is $\bar{y}_T = -16.073$ meters.

The height of sight scope of the rifle is $h_t = 2.7" = 006858m$.

(a) Determine the correction in aiming angle needed to shoot the center of target at $x_T = 1,200$ meters.

(b) Determine the control point at range $x_D = 100$ meters.

Solution

The marksman has setup the sight to zero the rifle at $x_D = 1,200$ meters using a departure angle of

$$\alpha_T = \frac{|\bar{y}_T|}{x_{TD}} \cdot \frac{180}{\pi} = \frac{16.073}{1200} \cdot \frac{180}{\pi} = 0.7674°.$$

The corresponding aiming angle is

$$\alpha_{ST} \approx \alpha_T + (\frac{h_T}{x_T}) \cdot \frac{180}{\pi} = (0.7674) + (\frac{0.06858}{1200}) \cdot \frac{180}{\pi} = 0.84774°.$$

(a) Employing (1.6.3), with control point $y_P = 0$, we have a change of

$$\Delta \alpha_T = (\frac{y_Q}{x_T}) \cdot \frac{180}{\pi} = \frac{0.5}{1200} \cdot \frac{180}{\pi} = 0.02387°$$

in departure angle.

The correct departure angle that really zeroes the rifle at $x_T = 1,200$ meters is

$$\alpha_{OT} = \alpha_T - \Delta \alpha_T = 0.7674 - 0.02387 = 0.7435°$$

The corresponding aiming angle is

$$\alpha_S \approx \alpha_T + (\frac{h_T}{x_T}) \cdot \frac{180}{\pi} = (0.7435) + (\frac{0.06858}{1200}) \cdot \frac{180}{\pi} = 0.8238°.$$

Since we keep unchanged the real departure angle that zeroes the rifle at $x_D = 1,200$, we have to change the sight-scope angle with the quantity

$$\Delta \alpha_h = \alpha_{S2} - \alpha_{SD} = 0.8238 - 0.84774 = -0.02387°.$$

(b) The control point at 100 meters is

$$y_P = x_{100} \cdot \alpha_{OD} \cdot \frac{\pi}{180} = 100 \cdot (0.7435°) \cdot \frac{\pi}{180} = 1.30 m$$

over the lower edge of the target board, located at 100 meters.
1,

Thus, at 100 meters the trajectory of the bullet must pass through the control point in order that the firearm should be zeroed at $x_T = 1,200$ meters.

1.7 Coordinate System Associated with Line of Sight

In practice of shooting, the marksman is interested to know the **height of a flying bullet** (trajectory height) above the line of sight (LOS) when the rifle is zeroed at a certain range $x_T = OT$ (fig. 2 above). In other words, the marksman wants to know the height $y' = EP$ of the trajectory at any range $x = OT$.

(Some exterior ballistic authors refer to the height of trajectory above or below the line of sight as **bullet path**.)

Note that for the range beyond the zero-range $x = OT$, the distance of the bullet from the line of sight is negative.

Let's show a simple way to estimate the height of trajectory over the line of sight. For this we introduce a coordinate system that has the x-axis along the LOS.

In exterior ballistics of long range shooting with small firearms the origin of the system of coordinates is located at the upper point S of the rifle scope, the x'-axis is directed along the line of sight (ST), while the y'-axis is perpendicular to the LOS. That system of coordinates is denoted (x'Sy').

Actually, the origin of the coordinate system (xoy), fig. 2 section 1.2, used in exterior ballistics to predict the projectile trajectory is located at the muzzle of the firearm; the direction of x-axis is along the horizon, while the y-axis is perpendicular to it.

Since the trajectory elements are obtained from the system of differential equations related with xoy system, we need to convert those data into the system x"Sy'.

Thus, for example, we need to convert the coordinates (x, y) of the trajectory projectile into the corresponding coordinates (x', y') relative to x'Sy'.

Let's find the relationship between those two coordinate systems (figure 2, section 1.2).

The system of coordinates x'Sy' is obtained from the system of coordinates xoy by shifting vertically x-axis and y-axis y_S units till the origin of coordinates will coincide with the point S and then rotating it clockwise with an angle equal to the angle of sight $\alpha_h = \angle OTS$.

The relationship between the coordinates of a point P(x, y) on the trajectory and the coordinates (x', y') of the same point P, but in the x'Sy' system, when rotation is clock-wise, is:

$$x' = x\cos(\alpha_h) - y\sin(\alpha_h),$$
$$y' = x\sin(\alpha_h) + y\cos(\alpha_h) - y_S, \qquad (1.7.1)$$

where $y_S = h$ is the y-coordinate of point S. The height of the scope is h.

ELEMENTS OF EXTERIOR BALLISTICS

The coordinates (x, y) of a point on the trajectory when are known the coordinates (x', y') are:

$$x = x'\cos(\alpha_h) + y'\sin(\alpha_h),$$
$$y = -x'\sin(\alpha_h) + y'\cos(\alpha_h) + y_S. \qquad (1.7.2)$$

The rotation angle is positive, i.e. $\alpha_h > 0$.

Since the angle $\alpha_h = \angle OTS$ is very small, it follows that $\cos\alpha_h \approx 1$ and $\sin\alpha_h \approx 0$. From (1.7.1) and (1.7.2), for the coordinates of the projectile, we have respectively:

$$x' = x,$$
$$y' = x\sin(\alpha_h) + y\cos(\alpha_h) - y_S, \qquad (1.7.3)$$

and

$$x = x',$$
$$y = -x'\sin(\alpha_h) + y'\cos(\alpha_h) + y_S, \qquad (1.7.4)$$

where $y_S = h > 0$.

Note that y' is the perpendicular distance of the point-mass bullet from LOS.

Illustration

(a) The coordinates of the projectile in (xoy) are respectively (500m, 1.5m), while the scope height is 1.5 inches (0.0381m).

The sight angle is

$$\alpha_h \approx \frac{h}{x} \cdot \frac{180}{\pi} = \frac{0.0381}{500} \cdot \frac{180}{\pi} = 0.004366°.$$

Using (1.7.3) we find:

$$x' = 500,$$
$$y' = 500 \cdot \sin(0.004366) + 1.5 \cdot \cos(0.004366) - 0.0381 = 1.424m,$$

i.e. that for the range 500 meters the height of the bullet over the line of sight is 1.462 meters.

(b) The coordinates of the projectile in (x'Sy') are respectively (500m, - 1,5m), while the scope height is 1.5 inches (0.0381m). Substituting in (1.7.4) we find:

$x = 500$,

$y = 500 \cdot \sin(0.004366) + (-1.5)\cos(0.004366) + 0.0381 = -1.424m$.

For the range 500 meters the bullet trajectory is 1.462 meters below the line.

Projectile Velocity and "LOS" Coordinate System

The projectile velocity v is not dependent on the system of coordinates, i.e. the projectile velocity, at any given point on the trajectory, is the same no matter which reference system of coordinates we use.

The above statement is true since from ((1.7.3) or (1.7.4) it follows that the components of velocity along x-axis and y-axis are equal, i.e.

$dx'/dt = dx/dt$ and $dy'/dt = dy/dt$.

The angle the velocity forms with the x-axis is different from the angle the velocity forms with the LOS, i.e. x'-axis (fig. 1).

Note that t is the temperature of propellant charge at the shooting point. In general, t is different from the temperature t_0 that represents the standard temperature of propellant charge (21.11 degree Celsius for the ICAO atmosphere, and ASM atmosphere, 15 degree Celsius for TSA).

In practice of shooting it is not always possible to store the bullet cartridge or propellant charge of cannon at standard temperature t_0).

Note that the line of sight Sx' is not a fixed coordinate axis. It's direction depends on the location of aiming point, for example T or D.

Thus, the point D located on the ox axis that passes from the origin of coordinates 0 through T is below the LOS.

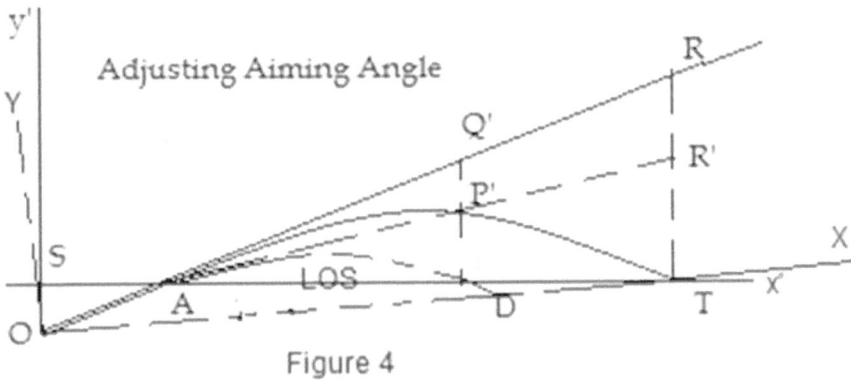

Figure 4

1.8 Height of Trajectory over Line of Sight

We assume that the firearm is zeroed at the horizontal range $x_T = OT$. The trajectory of bullet (fig. 2) passes at the point P which is above LOS, when the zeroing range ($x_T = OT$) is greater than $x_D = OD$, or below LOS, when the zeroing range, $x_T = OT$, is smaller than $x_D = OD$).

The height y'_P of the trajectory above LOS can be found using equation (1.7.1), i.e.

$$x'_D = x_D,$$
$$y'_D = x_D \sin(\alpha_h) + y_D \cos(\alpha_h) - h, \tag{1.8.1}$$

where

$$y_D = x_D (\tan \alpha_{0T} - \tan \alpha_D), \tag{1.8.2}$$

or

$$y_D = x_D \left(\frac{|\bar{y}_T|}{x_T} - \frac{|\bar{y}_D|}{x_D} \right), \tag{1.8.3}$$

where α_{0T} is the departure angle that zeroes the firearm at the horizontal range, $x_T = OT$, α_D is the angle that zeroes the firearm at range $x_D = OD$, $\bar{y}_T$ and $\bar{y}_D$ are the bullet drops respectively at ranges $x_T = OT$ and $x_D = OD$, $y_S = h$ is the scope height.

When it is known the height y'_P of bullet trajectory above (below) the line of sight, using (1.7.4) we find that the height y_D of the bullet above the horizontal line is

$$x_D = x'_D,$$
$$y_D = -x'_D \sin(\alpha_h) + y'_D \cos(\alpha_h) + h. \quad (1.8.4)$$

The following examples demonstrate the application of above formulae to find the height of trajectory above/below the line of sight, or the height of trajectory over/under the horizontal line.

Example 8.1 Lapua bullet, metric units

Refer to table 3.1, section 5.3.

Assume that the rifle is zeroed at $x_6 = 600m$. The departure angle that corresponds to the zero range $x_6 = 600m$ is $\alpha_{0T} = 0.3018°$.

Find the height y'_P of the bullet trajectory above/below the line of sight for the range $x_3 = 300m$ and $x_{10} = 1,000m$, if the sight height (scope above the bore line) is $h = 0.04m$.

The departure angle that corresponds to the zero-ranges: $x_3 = 300m$, $x_6 = 600m$ and are respectively: $\alpha_3 = 0.1353°$ and $\alpha_{10} = 0.5958°$.

Solution

At $x_6 = 600$ meters, the angle of sight-scope, in absolute value, is

$$\alpha_h = \frac{h}{x} \cdot \frac{180}{\pi} = \frac{0.040}{600} \cdot \frac{180}{\pi} = 0.00382°.$$

(a) At $x_3 = 300m$, the height above the ground of the bullet trajectory (rifle zeroed at $x_6 = 600$ meters) is

$$y_D = x_D \cdot (\tan \alpha_0 - \tan \alpha_D) =$$

$$= 300 \cdot [\tan(0.3018) - \tan(0.1353)]) = 0.872m$$

Substituting in (1.8.1), we find that the height of bullet over the LOS is

$$y'_D = x_D \sin(\alpha_h) + y_D \cos(\alpha_h) - h =$$
$$= 300 \cdot \sin(0.00382) + 0.872 \cdot \cos(0.00382) - 0.040 = 0.852 m.$$

This result is equal to the value 0.855 meters shown by Lapua[3] for the given bullet.

(b) At $x_D = 1,000$ meters, the depth of the trajectory below horizontal line is
$$y_D = 1,000 \cdot (\tan(0.3018) - \tan(0.5958)) = -5.1316 m$$

The distance of bullet trajectory from LOS is
$$y'_D = 1,000 \cdot \sin(0.00382) + (-5.1316) \cdot \cos(0.00382) - 0.040$$
$$= -5.105 m.$$

The above result is quite equal to the value -5.082 shown by Lapua Company[4].

Example 8.2 Lapua bullet, imperial units

The departure angle needed to hit the target at the zero-range equal to $x_T = 1,000 \ yards$, when the rifle fires a 0.338 Lapua GB528 Scenar 19.44g bullet, in ICAO atmosphere, is $\alpha_0 = 0.5247°$.

For the range $x_D = 800 \ yards$, find the height of the bullet trajectory above LOS if the scope above the bore line is $y_S = 1.6 \ inches$.

The departure angle that corresponds to the zero-range $x_D = 800 \ yards$ is $\alpha_8 = 0.3880°$.

Solution

The sight-scope angle is
$$\alpha_h = \frac{h}{x} \cdot \frac{180}{\pi} = \frac{1.60}{1000 \cdot (36)} \cdot \frac{180}{\pi} = 0.00255°.$$

At range $x_D = 800$ yards, the height of the trajectory above the ground (ox axis) is
$$y_D = 36 \cdot x \cdot (\tan \alpha_{10} - \tan \alpha_8) = 36 \cdot 800 \cdot (\tan(0.5247)$$
$$- \tan(0.3880)) = 68.717 in$$

For the range 800 yards, using the second formula of (8), we find that the height of the trajectory of bullet above the line of sight is

$$y'_D = 800 \cdot (36) \cdot \sin(0.00255) + 68.717 \cdot \cos(0.00255)$$
$$-1.60 = 68.40 in.$$

Example 8.3. Height of trajectory over horizontal line

The rifle firing a 0.338 Lapua GB528 Scenar 19.44g bullet is zeroed at the range 600 yards.

According to Lapua range table (see end note 3), at range 1,000 yards, the height of the trajectory over the LOS is -158 inches.

Find the height of the projectile below the horizontal line.

The sight height is 1.6 inches.

Solution

The sight-scope angle is

$$\alpha_h = \frac{h}{x} \cdot \frac{180}{\pi} = \frac{1.60}{1000 \cdot (36)} \cdot \frac{180}{\pi} = 0.00255°.$$

Substituting in (1.8.4) we have:

$$x = x' = 1,000 \text{ .yards}$$
$$y = -x'\sin(\alpha_h) + y'\cos(\alpha_h) + h =$$
$$= -1000 \cdot 36 \cdot \sin(0.00255) + (-158) \cdot \cos(0.00255) + 1.6 = -158 in.$$

1.9 Change of Zero Range Using Bullet Path

Assume that a firearm is zeroed in at a range $x_T = OT$ (figure 2, figure 4). The sight scope height is h. We need to adjust the aiming angle in order to change the zeroing in at the new range $x_D = OD$.

The range $x_D = OD$ can be smaller or greater than $x_T = OT$.

Assume as well that we know the height $y'_D = DP'$ (fig. 2.) of the trajectory above or below the horizontal line (bullet path).

The aiming angle at $x_D = OD$ is

$$\alpha_{SD} = \frac{y'_D}{x'_D} \cdot \frac{180}{\pi}. \tag{1.9.1}$$

From figure 2, we can see that:

$$y'_D \approx x'_D (\alpha_{ST} - \alpha_{SD}) \cdot \frac{\pi}{180}, \tag{1.9.2}$$

where α_{ST} and α_{SD} are the aiming angles corresponding respectively to $x'_D = OD$ and $x_T = OT$.

Since $x'_D = x_D$ and $y'_D = DP'$, we can write:

$$y'_D \approx x_D (\alpha_{ST} - \alpha_{SD}) \cdot \frac{\pi}{180}. \tag{1.9.3}$$

Hence, for the change in aiming angle, when we change the zeroing in from $x_T = OT$ to $x_D = OD$ we have:

$$\Delta \alpha_S = (\alpha_{SD} - \alpha_{ST}) = -\frac{y'_D}{x_D} \cdot \frac{180}{\pi}. \tag{1.9.4}$$

The change in angle of sight-scope is

$$\Delta \alpha_h = \Delta \alpha_S. \tag{1.9.5}$$

Substituting (1.9.4) into (1.9.5) we find that the change in angle of sight scope is

$$\Delta \alpha_h = -\frac{y'_D}{x_D} \cdot \frac{180}{\pi}. \tag{1.9.6}$$

Example 9.1 Method 1, Sight Adjustment

A rifle firing a 0.308 Sierra 165 grain Spitzer boattail bullet, with velocity 2,700 fps, is zeroed at the range $x_T = 250$ yards.

According to Sierra, at zero range $x_T = 250$ yards the bullet drop is -16.61 inches.

(a) Find angle of sight-scope.

(b) For the range 100 yards, find the height of the bullet trajectory above horizontal line.

(c) For the range 100 yards, find the height of bullet trajectory over LOS, if the height of sight scope above the bore line is $y_S = 1.8 \ inches$.

d) Find the sight adjustment when we change the zeroing from 250 yards into 100 yards.

Solution

(a) The sight-scope angle, in absolute value, at $x_T = 250$ yards is

$$\alpha_h = \frac{h}{x_T} \cdot \frac{180}{\pi} = \frac{1.80}{250 \cdot (36)} \cdot \frac{180}{\pi} = 0.01146°.$$

(b) Substituting in (1.8.3), we find that the height of bullet over the horizontal line at 100 yards is

$$y_D = x_D \left(\frac{|\bar{y}_T|}{x_T} - \frac{|\bar{y}_D|}{x_D} \right) = 100 \cdot \left(\frac{16.61}{250} - \frac{2.37}{100} \right) = 4.274 in.$$

(c) At 100 yards the height of bullet over LOS is

$$y'_D = 100 \cdot (36) \cdot \sin(0.01146) + 4.274 \cdot \cos(0.01146) - 1.80,$$
$$= 3.194 in$$

i.e. at 100 yards the bullet will hit 3.194 inches above the center of the target (aiming point), considering that the rifle is zeroed at $x_T = 250$ yards.

The corresponding change in angle of sight-scope is

$$\Delta \alpha_h = \frac{y'_D}{x_D} \cdot \frac{180}{\pi} = \frac{3.194}{100 \cdot (36)} \cdot \frac{180}{\pi} = 0.05073° = 3.05 MOA.$$

Example 9.2 Change of Zero Range. Method 1

The departure angle needed to hit the target at the zero-range equal to $x_T = 600$ yards, when the rifle fires a 0.338 Lapua GB528 Scenar 19.44g bullet, in ICAO atmosphere, is $\alpha_{0T} = 0.2709°$ (velocity 2723.10 fps).

For the range $x_D = 900$ yards, find the height of the bullet trajectory below LOS if the scope above the bore line is $y_S = 1.6 \ inches$.

The departure angle that corresponds to the zero-range $x_D = 900$ yards is $\alpha_{0D} = 0.4535°$.

Solution

The angle of sight-scope at $x_T = 600$ yards is

$$\alpha_h = \frac{h}{x_T} \cdot \frac{180}{\pi} = \frac{1.60}{600 \cdot (36)} \cdot \frac{180}{\pi} = 0.00424°.$$

The height of bullet below the horizontal line at $x_D = 900$ yards is

$$y_D \approx x_D (\alpha_{0T} - \alpha_{0D}) \cdot \frac{\pi}{180} = 900 \cdot (36) \cdot (0.2709 - 0.4535) \cdot \frac{\pi}{180}$$

$$= -103.26 in.$$

At 900 yards the depth of bullet below LOS is

$$y'_D = 900 \cdot (36) \cdot \sin(0.00424) + (-103.26) \cdot \cos(0.00424) - 1.60,$$

$$= -102.464 in$$

The corresponding change in angle of sight-scope is

$$\Delta \alpha_h = \frac{y'_D}{x_D} \cdot \frac{180}{\pi} = \frac{-102.46}{900 \cdot (36)} \cdot \frac{180}{\pi} = -0.1812° = -10.87 MOA.$$

1.10 Change of Zero Range Using Trajectory Height

Assume that a firearm is zeroed in at a range $x_T = OT$. The sight scope height is h. We need to adjust the aiming angle in order to change the zeroing in at the new range $x_D = OD$.

The range $x_D = OD$ can be smaller or greater than $x_T = OT$.

Control point P corresponds to the shooting in standard conditions when the zero range is at $x_T = OT$ (figure 4).

Assume we know the bullet path $y'_D = DP'$ at range x_D when the zero range in standard conditions is $x_T = OT$.

As result of test shooting the center of group hits is $y'_Q = DQ'$. The aiming angle that correspond to x_D, and to bullet path $y'_D = DP'$ is

$$\alpha_{SP'} = \frac{y'_D}{x_D} \cdot \frac{180}{\pi}. \tag{1.10.1}$$

The aiming angle that correspond to x_D, and to bullet path $y'_Q = DQ'$ is

$$\alpha_{SQ'} = \frac{y'_Q}{x_D} \cdot \frac{180}{\pi}. \tag{1.10.2}$$

The change in aiming angle is

$$\Delta\alpha_S = (\frac{y'_{Q'} - y'_{P'}}{x_D}) \cdot \frac{180}{\pi}. \tag{1.10.3}$$

Thus, for the change in sight scope angle we have:

$$\Delta\alpha_h = (\frac{y'_{Q'} - y'_{P'}}{x_D}) \cdot \frac{180}{\pi}. \tag{1.10.4}$$

Let's simplify (1.10.3).

According to equation (1.8.1), we can express the bullet path through the height of trajectory over the horizontal line, i.e. we can write:

$$y'_P = x_D \sin(\alpha_h) + y_D \cos(\alpha_h) - h \tag{1.10.5}$$

and

$$y'_Q = x_D \sin(\alpha_h) + y_Q \cos(\alpha_h) - h. \tag{1.10.6}$$

Substituting in (1.10.3), we find that

$$\Delta\alpha_S = (\frac{y_Q - y_P}{x_D}) \cdot \cos\alpha_h \cdot \frac{180}{\pi}, \tag{1.10.7}$$

where the sight-scope angle is

$$\alpha_h = -\frac{h}{x_T} \cdot \frac{180}{\pi}. \tag{1.10.8}$$

Thus we can write:

$$\Delta\alpha_h = (\frac{y_Q - y_D}{x_D}) \cdot \cos\alpha_h \cdot \frac{180}{\pi}, \tag{1.10.9}$$

We see that there is an irrelevant difference between (1.10.9) and (1.10.4) since $\cos\alpha_h \approx 1$.

We have to write (1.10.9) in the following form:

$$\Delta\alpha_h = (\frac{y_Q - y_D}{x_D}) \cdot \cos\alpha_h \cdot \frac{180}{\pi}. \tag{1.10.10}$$

Example 10.1

A competitive shooter fires some 0.338 Lapua GB528 Scenar 19.44g bullet with a departure speed of 830 m/s on a control table located 100 meters from the rifle.

Using the standard range table 3.1, section 5.3, the shooter finds that the departure angle that zeroes the rifle at $x_T = 500$ meters is $\alpha_{0T} = 0.242°$, while the departure angle that zeroes the rifle at $x_D = 100$ meters is $\alpha_{OD} = 0.0419°$.

The control point P on the firing board, i.e. the height of the bullet trajectory above the horizontal line at $x_D = 100$ meters is 0.349.

Using ballistic standard tables, the shooter sets up the aiming angle in order to zero the rifle at $x_T = 500$ meters.

To test the scope set up, the shooter fires at a control board located at $x_D = 100$ meters.

The actual center of the group of shoots is 0.360 meters above the horizontal line.

- Adjust the sight height to zero the firearm at $x_T = 500$.
- What is the height of the bullet trajectory at $x_T = 500$ at the actual conditions?

Solution

The sight scope is not set up correctly to zero the firearm at $x_T = 500$. The sight scope correction is

$$\Delta\alpha_h = -(\frac{y_Q - y_D}{x_D}) \cdot \cos\alpha_h \cdot \frac{180}{\pi} =$$

$$= -(\frac{0.360 - 0.349}{100}) \cdot \cos(0.00458) \cdot \frac{180}{\pi} = 0.0063°.$$

At $x_T = 500$ the bullet height is (see fig 4):

$$y_T = x_T \cdot \Delta\alpha_{0T} = x_T \cdot (\frac{y_Q - y_D}{x_D}) = 500 \cdot \frac{0.36 - 0.349}{100} = 0.055m$$

above the center of the target.

1.11 Trajectory Height and Bullet Path Tables

The standard range tables do not include the height of the trajectory above the line of sight (LOS), i.e. bullet path.

Using standard range tables and the equations obtained in above sections, we will show the way we can construct the Trajectory Path Tables using standard range tables.

For illustration we are going to construct the path table for the 0.338 Lapua GB528 Scenar bullet 19.44g (initial velocity 830 m/s, ICAO atmosphere), using the Standard Range Table 3.1 section 5.3.

Scope height is $h = 0.040m$.

(a) **We assume that the rifle is zeroed at $x_T = 600$ meters** (departure angle $\alpha_{0T} = 0.3018°$. Let's find the height of bullet trajectory above the LOS for ranges 100 meters till 700 meters.

Using 10.1, we find that the angle of scope-sight is

$$\alpha_h = \frac{h}{600} \cdot \frac{180}{\pi} = 0.0038°.$$

(1) Range $x_D = 100m$. Departure angle $\alpha_{0D} = 0.0419°$.

Using (1T), we find the height of the trajectory over horizontal line:

$$y_D = x_D(\alpha_{0T} - \alpha_{0D}) \cdot \frac{\pi}{180} = 100 \cdot (0.3018 - 0.0419) \cdot \frac{\pi}{180} = 0.4536m.$$

The height of trajectory over LOS (the path) is:
$$y'_D = x_D \sin(\alpha_h) + y_D \cos(\alpha_h) - y_S =$$

$$= 100 \cdot \sin(0.0038) + 0.4536 \cdot \cos(0.0038) - 0.040 = 0.420m.$$

(2) Range $x_D = 200m$, departure angle $\alpha_{0D} = 0.0691°$.

We have:
$$y_D = x_D(\alpha_{0T} - \alpha_{0D}) \cdot \frac{\pi}{180} = 200 \cdot (0.3018 - 0.0691) \cdot \frac{\pi}{180}$$

$$= 0.8123m.$$

$$y'_D = x_D \sin(\alpha_h) + y_D \cos(\alpha_h) - y_S =$$

$$= 200 \cdot \sin(0.0038) + 0.8123 \cdot \cos(0.0038) - 0.040 = 0.786m$$

(7) Range $x_D = 700m$, departure angle $\alpha_{0D} = 0.3662°$.

We have:
$$y_D = x_D(\alpha_{0T} - \alpha_{0D}) \cdot \frac{\pi}{180} = 700 \cdot (0.3018 - 0.3662) \cdot \frac{\pi}{180}$$

$$= -0.7868m$$

$$y'_D = x_D \sin(\alpha_h) + y_D \cos(\alpha_h) - y_S =$$

$$= 700 \cdot \sin(0.0038) - 0.7868 \cdot \cos(0.0038) - 0.040 = -0.7801m$$

In the same way we find the height of bullet trajectory for other ranges.

(b) **We assume that the rifle is zeroed at $x_T = 700$ meters** (departure angle $\alpha_{0T} = 0.3662°$. Let's find the height of bullet trajectory above the LOS for ranges 100 meters till 800 meters.

Using 10.1, we find that the angle of scope-sight is
$$\alpha_h = \frac{h}{700} \cdot \frac{180}{\pi} = 0.00327°.$$

(6) Range $x_D = 600m$. Departure angle $\alpha_{0D} = 0.3018°$.

Using (1T), we find the height of the trajectory over horizontal line:
$$y_D = x_D(\alpha_{0T} - \alpha_{0D}) \cdot \frac{\pi}{180} = 600 \cdot (0.3662 - 0.3018) \cdot \frac{\pi}{180}$$
$$= 0.6744m.$$

The height (path) of trajectory over LOS is
$$y'_D = x_D \sin(\alpha_h) + y_D \cos(\alpha_h) - y_S =$$
$$= 600 \cdot \sin(0.00327) + 0.6744 \cdot \cos(0.00327) - 0.040 = 0.6687m.$$

Using the described procedure we have constructed the folllowing tables:

- Trajectory Height over the Horizontal Line Table 11.1.
- Trajectory Height over LOS Table 11.2.

Table 11.1 Trajectory Height: 0.338 Lapua Magnum Scenar GB528, 19.4g

Rifle Sighted	Trajectory Height Above Horizontal Line, [Meters]									
	100	200	300	400	500	600	700	800	900	1000
100	0.000	-0.095	-0.489	-1.008	-1.746	-2.721				
200	0.048	0.000	-0.347	-0.818	-1.508	-2.437				
300	0.163	0.231	0.000	-0.356	-0.931	-1.744	-2.821			
400	0.252	0.409	0.267	0.000	-0.486	-1.210	-2.198	-3.497		
500	0.349	0.604	0.559	0.389	0.000	-0.626	-1.517	-2.718	-4.251	
600	0.454	0.812	0.872	0.806	0.522	0.000	-0.787	-1.884	-3.31	
700	0.566	1.037	1.209	1.256	1.084	0.674	0.000	-0.980	-2.300	
800	0.68	1.283	1.578	1.748	1.699	1.413	0.861	0.000	-1.19	-2.78
900	0.822	1.548	1.976	2.278	2.361	2.207	1.789	1.060	0.000	-1.452
1,000	0.967	1.839	2.41.3.3	2.859	3.087	3.079	2.805	2.221	1.307	0.000

Table 11.2 Path: 0.338 Lapua Magnum Scenar GB528, 19.4g

Rifle Sighted	Trajectory height Above LOS, [Meters]. Scope Height 0.040m									
	100	200	300	400	500	600	700	800	900	1000
100	0.000	-.0549	-.409	-0.888	-1.586	-2.521				
200	0.027	0.000	-0.327	-0.778	-1.449	-2.357				
300	0.136	0.218	0.000	-0.343	-0.904	-1.704	-2.768			
400	0.222	0.389	0.257	0.000	0.476	-1.190	-2.168	-3.456		
500	0.317	0.580	0.543	0.381	0.000	0.618	-1.501	-2.695	-4.219	
600	0.420	0.786	0.852	0.793	0.515	0.000	-0.780	-1.870	-3.291	
700	0.532	1.009	1.186	1.256	1.072	0.669	0.000	-0.979	-2.290	
800	0.654	1.253	1.553	1.728	1.684	1.403	0.856	0.000	-1.19	-2.70
900	0.786	1.517	1.949	2.256	2.344	2.194	1.780	1.055	0.000	-1.45
1,000	0.931	1.807	2.383	2.835	3.067	3.063	2.793	2.213	1.303	0.000

Both tables can be used to change the zeroing in, employing equations (1.5.15),

$$\Delta\alpha_{hT} = (-\frac{y_D}{x_D} + \frac{h_T}{x_D} - \frac{h_T}{x_T}) \cdot \frac{180}{\pi}$$

and (1.9.6)

$$\Delta\alpha_{hT} = -\frac{y'_D}{x_D} \cdot \frac{180}{\pi}.$$

The method 1 that uses equation (1.5.15) and table 11.1 can be used to change the zeroing in even when the height of sight scope of the rifle is not the sight-scope height predetermined to construct table 11.2 ($h_T = 0.04$ meters in table 11.2).

The illustrations in example 11.1 and example 11.2, show the way we find the change in sight-scope angle.

The tables can be used to determine the Point Blank Range, etc.

Example 11.1 Trajectory Height and Change of Sight

Use the data in table 11.1 to change the zeroing from $x_T = 600$ meters to $x_D = 500$ meters, when the height of the sight-scope is $h_T = 0.040$ meters.

Solution

Given that the rifle is zeroed at $x_T = 600$ meters, then at $x_D = 500$ meters the trajectory height over the horizontal line is $y_D = 0.522$ meters (table 11.1).

Employing (1.5.15), we find that the change in sight-scope angle is

$$\Delta\alpha_h = (-\frac{y_D}{x_D} + \frac{h_T}{x_D} - \frac{h_T}{x_T}) \cdot \frac{180}{\pi} =$$

$$= (-\frac{0.522}{500} + \frac{0.040}{500} - \frac{0.040}{600}) \cdot \frac{180}{\pi} = -0.05905°.$$

Example 11.2 Trajectory Height and Change of Sight

Use the data in table 11.1 to change the zeroing from $x_T = 600$ meters to $x_D = 500$ meters, when the height of the sight-scope is $h_T = 0.06858$ meters.

Solution

When rifle is zeroed at $x_T = 600$ meters, then at $x_D = 500$ meters the trajectory height over the horizontal line is $y_D = 0.522$ meters (table 11.1).

Employing (1.5.15), we have

$$\Delta\alpha_h = (-\frac{y_D}{x_D} + \frac{h_T}{x_D} - \frac{h_T}{x_T}) \cdot \frac{180}{\pi} =$$

$$= (-\frac{0.522}{500} + \frac{0.06858}{500} - \frac{0.6858}{600}) \cdot \frac{180}{\pi} = -0.05851.$$

Example 11.3 Bullet Path and Change of sight

Use the data of table 11.2 to change the zeroing in from 600 meters to 500 meters.

Solution

When rifle is zeroed at $x_T = 600m$, we find that the path of the trajectory over LOS at $x_D = 500$ meters is $y'_D = 0.515m$ table 11.2).

Substituting in (1.9.6), we find that the change in sight scope angle is

$$\Delta\alpha_h = -\frac{y'_D}{x_D} \cdot \frac{180}{\pi} = -\frac{0.515}{500} \cdot \frac{180}{\pi} = -0.05905° = -3.541 MOA.$$

Note that both methods, for the same sight scope height, give equal changes in sight-scope height.

2

Long Range Inclined Shooting

Introduction

In this chapter we show some simple and practical methods to estimate the super elevation or depression angle in uphill, or downhill long range shootings. The methods are approximate but incredibly accurate.

We also elaborate mathematically some of the exterior ballistics definitions, concepts, principles, formulas, etc., such as the rigidity principle and non-rigidity principle of the projectile trajectory.

We introduce "The Non-Rigidity-Principle" model that can be used to find the super elevation or super depression angle and to set up the aiming angle in uphill or downhill shooting.

The non-rigidity principle is used in quite all my books without mentioning it.

We consider the point-mass projectile trajectory without taking into account the corrections for the wind, rotation of bullet, rotation of Earth, etc.

In practice of long range shooting, to make feasible the process of shooting in the field of fast moving and hiding targets, the rifleman ignores many of the above factors that influence the accuracy of firing.

It is in the shooter discretion and in his/her shooting preparation to ignore or consider one or the other factor that influences the accuracy in long range shooting.

At the end of the chapter there are included some exercises that are solved using PC programs associated with the book. Through those exercises we verify the accuracy of approximate methods we use to predict the projectile trajectory, for example the approximate formulas related with inclined shooting.

2.1 A Practical Method on Inclined Shooting

Super Elevation Angle

A simple and accurate approach to estimate the super elevation angle during uphill or downhill shooting is based on the following formula[5]:

$$\alpha_{OI} = \alpha_{OT} \cdot \cos(E), \qquad (2.1.1)$$

where :

- $\alpha_{OT} = \angle TOT_1$ is the departure angle (in degree) that corresponds to the horizontal range $x_T = OT$ (in yards, or meters),
- $E = R_x R$ is the elevation or depression angle (in degree),
- $\alpha_{OI} = \angle ROR_1$ is the super elevation (super depression) angle that corresponds to the slant range $R = OR$, which is equal to the horizontal range $x_T = OT$, i.e. $R = x_T$ (fig. 5).

Note that the inclined range is equal to the corresponding horizontal range, i.e. $R = x_T$.

The horizontal range can be seen as an inclined shooting with elevation or depression angle $E = 0$.

Inclined Aiming Angle

The angle of sight α_S (in degree), for the inclined or declined shooting, can be estimated using the following formula:

$$\alpha_S = \alpha_{OI} - \alpha_h, \qquad (2.1.2)$$

where α_h is the sight-scope angle ($\alpha_h < 0$).

Since the sight-scope angle that zeroes in the firearm at the inclined range $R = OR = x_T$ is

$$\alpha_h = -\frac{h}{R} \cdot \frac{180}{\pi}, \qquad (2.1.3)$$

Since $R = x_T$, we e can write for the aiming angle:

$$\alpha_S = \alpha_{OI} + \frac{h}{x_T} \cdot \frac{180}{\pi}, \qquad (2.1.4)$$

where h is the sight-scope height.

For the uphill shooting the elevation angle E is positive, while for the downhill shooting, the depression angle E is negative.

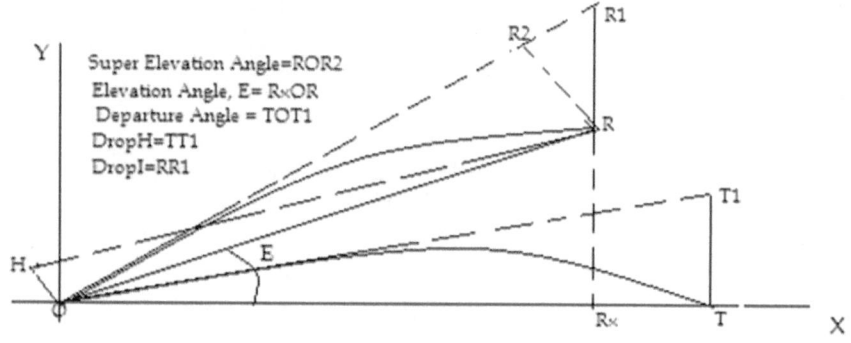

"Figure 5"

For Lapua Scenar GB528, 19.44g bullet, in ICAO atmosphere, the super elevation angle (estimated using equation (2.1.1)) for different inclined ranges is given in table 1.1.

Table 1.1. Super Elevation Angle, 0.308 Lapua Scenar GB528, 19.44g Bullet, ICAO Atmosphere

Range (yards)	Elevation Angle (Degree)									
	0	15	20	25	30	35	40	45	50	55	60
	Super Elevation Angle (Degree)									
100	0.0379	0.0366	0.0356	0.0343	0.0328	0.0310	0.0290	0.0268	0.0244	0.0217	0.0190
200	0.0789	0.0762	0.0741	0.0715	0.0683	0.0646	0.0604	0.0558	0.0507	0.0453	0.0395
300	0.1218	0.1176	0.1145	0.1104	0.1055	0.0998	0.0933	0.0861	0.0783	0.0699	0.0609
400	0.1688	0.1630	0.1586	0.1530	0.1462	0.1383	0.1293	0.1194	0.1085	0.0968	0.0844
500	0.2178	0.2104	0.2047	0.1974	0.1886	0.1784	0.1668	0.1540	0.1400	0.1249	0.1089
600	0.2709	0.2617	0.2546	0.2455	0.2346	0.2219	0.2075	0.1916	0.1741	0.1554	0.1355
700	0.3273	0.3161	0.3076	0.2966	0.2835	0.2681	0.2507	0.2314	0.2104	0.1877	0.1637
800	0.3877	0.3745	0.3643	0.3514	0.3358	0.3176	0.2970	0.2741	0.2492	0.2224	0.1939
900	0.4535	0.4380	0.4262	0.4110	0.3927	0.3715	0.3474	0.3207	0.2915	0.2601	0.2268
1000	0.5247	0.5068	0.4931	0.4755	0.4544	0.4298	0.4019	0.3710	0.3373	0.3010	0.2624
1100	0.6005	0.5800	0.5643	0.5442	0.5200	0.4919	0.4600	0.4246	0.3860	0.3444	0.3003
1200	0.6831	0.6598	0.6419	0.6191	0.5916	0.5596	0.5233	0.4830	0.4391	0.3918	0.3416
1300	0.7731	0.7467	0.7264	0.7006	0.6695	0.6333	0.5922	0.5466	0.4969	0.4434	0.3865
1400	0.8698	0.8402	0.8173	0.7883	0.7533	0.7125	0.6663	0.6150	0.5591	0.4989	0.4349
1500	0.9758	0.9426	0.9170	0.8844	0.8451	0.7993	0.7475	0.6900	0.6272	0.5597	0.4879
1600	1.0913	1.0541	1.0255	0.9891	0.9451	0.8939	0.8360	0.7717	0.7015	0.6259	0.5457

In the second column of the table 1.1 is given the departure angle α_0 that corresponds to a given horizontal range (100 - 1600 yards), i.e. when elevation angle E is zero.

The estimation of the departure angle α_{0T} (elevation $E = 0$), presented in the second column, is done using PC program RLAPUA16.BAS (Appendix D).

Example 1.1 illustrates the way we have estimated the super-elevation angle and the data presented in table 1.1.

Example 1.1 Inclined Departure Angle

Using equation (2.1.1) let's find the super elevation angle α_{OI} for the 0.338 Lapua Scenar GB528 19.44 g (300 gr) 8.59 mm bullet, when the inclined range and the elevation angle are respectively $R_S = 1200$ yards and $E = 35°$.

The departure angle for the horizontal range $x_T = 1200$ yards (elevation angle $E = 0°$) is $\alpha_0 = 0.6831°$ (see table 1.1).

Solution

Substituting in formula (2.1.1) we find that the super elevation angle is

$$\alpha_{OI} = 0.6831° \cdot \cos(35°) = 0.5896°.$$

Example 1.2

For 0.338 Lapua GB528 bullet, find the super depression angle needed to hit a target located on an inclined plane 1200 yards from a sniper. The elevation angle is - 25 degree.

In the first column of table 1.1, for the horizontal range 1200 yards, we find that $\alpha_0 = 0.6831°$.

Solution

Substituting in (2.1.1) we find that the super elevation angle is

$$\alpha_{OI} = \alpha_{OT} \cdot \cos(E) = 0.6831 \cdot \cos(-25) = 0.6191°.$$

Example 1.3 Change of Zeroing in Horizontal Shooting

0.388 Lapua GB528 19.44 bullet.

Assume that the rifle is zeroed at the horizontal range 100 yards. Elevation angle $E = 0$. The sight height is $h = 2.7\ inches$.

(a) Find the aiming angle α_S that zeroes the rifle at the horizontal range $R_S = 100$ yards.

(b) Find as well the sight angle α_S that zeroes the rifle at the horizontal range 1000 yards.

Solution

From table 1.1 (elevation angle $E = 0$) we find that the departure angle that corresponds to horizontal range 100 yards is: $\alpha_0 = 0.0379°$.

In table 1.1 (elevation angle $E = 0$) we see that the departure angle that corresponds to horizontal range 1000 yards is: $\alpha_0 = 0.5247$.

(a) Horizontal range 100 yards

Substituting in (2.1.4) we find that the aiming angle is

$$\alpha_{S1} = \alpha_{OI} + \frac{h}{x_T} \cdot \frac{180}{\pi} = 0.0379 + \frac{2.7}{100 \cdot (36)} \cdot \frac{180}{\pi} = 0.08087°.$$

(b) Horizontal range 1000 yards: Substituting in (2.1.4) we find that the aiming angle is:

$$\alpha_{S2} = \alpha_{OI} + \frac{h}{x_T} \cdot \frac{180}{\pi} = 0.5247 + \frac{2.7}{1,000 \cdot (36)} \cdot \frac{180}{\pi} = 0.5290°.$$

Change in aiming angle is
$$\Delta\alpha_S = \alpha_{S2} - \alpha_{S1} = 0.5290° - 0.08087° = 0.4481° = 26.888 MOA.$$

Change in sight-scope angle is
$$\Delta\alpha_h = \Delta\alpha_S = 26.888 MOA.$$

Thus, we have to increase the sight scope angle with 26.888 MOA.

Number of Clicks

If we assume that 1 sight click is $1/8 = 0.125\ MOA$, we find that the number of clicks to change zeroing in from 100 yards to 1000 yards is
$$\Delta N = 26.888 / 0.125 = 215.10 \approx 215 clicks.$$

Example 1.4 Change in Zeroing in Inclined Shooting

Find the inclined aiming angle α_S needed to zero the rifle respectively at the inclined range $R_{S1} = 100$ yards and $R_{S2} = 1,000$ yards if the elevation angle is $E = 30$ degree.

Solution

In table 1.1 (elevation angle $E = 30°$), for range 100 yards we see that the super elevation angle is $\alpha_{OI} = 0.0328°$.

Substituting in (2.1.4) we find that the corresponding aiming angle is

$$\alpha_{S1} = \alpha_{OI} + \frac{h}{R_S} \cdot \frac{180}{\pi} = 0.0328 + \frac{2.7}{100 \cdot (36)} \cdot \frac{180}{\pi} = 0.07577°.$$

In table 1.1 (elevation angle $E = 30°$), for horizontal range 1,000 yards we see that the super elevation angle is $\alpha_{OI} = 0.4544$ degree.

Using (2.1.4) we find that the aiming angle is

$$\alpha_{S2} = \alpha_{OI} + \frac{h}{R_S} \cdot \frac{180}{\pi} = 0.4544 + \frac{2.7}{1,000 \cdot (36)} \cdot \frac{180}{\pi} = 0.4587°.$$

Change in aiming angle is

$$\Delta\alpha_S = \alpha_{S2} - \alpha_{S1} = 0.4587° - 0.07577° = 0.3829° = 22.976 MOA.$$

Change in sight-scope angle is

$$\Delta\alpha_h = \Delta\alpha_S = 22.976 MOA.$$

Number of Sight-Scope Clicks

If we assume that 1 sight click is $1/8 = 0.125$ MOA, we find that the number of clicks to zero the rifle at the horizontal range 1000 yards is

$$\Delta N = 22.976/0.125 = 183.81 \approx 184 clicks.$$

Example 1.5 Change of Zeroing in Inclined Shooting

Use the results obtained in example 1.2 and example 1.3 to estimate the change in aiming angle and in sight clicks when we change the zeroing in

(a) From the zero range 100 yards when the elevation angle is E = 0 to the zero range 100 yards when the elevation angle is E = 30 degree.

(b) From the zero range 1,000 yards when elevation angle is E = 0 to the zero range 1,000 yards when the elevation angle is E = 30 degree.

Solution

(a) Using the data in above examples we find that at 100 yards the change in aiming angle is

$$\Delta\alpha_S = \alpha_{S30} - \alpha_{S0} = 0.07577° - 0.08087° = -0.0051°$$
$$= -0.306 MOA.$$

Change in angle of sight-scope is

$$\Delta\alpha_h = \Delta\alpha_S = -0.306 = -0.306 MOA.$$

The number of sight clicks is

$$\Delta N = 0.306/0.125 = 2.45 \approx 2 clicks.$$

We have to reduce the sight scope clicks by 2 clicks.

(b) In the same way for 1,000 yards we find:
$$\Delta\alpha_S = \alpha_{S30} - \alpha_{S0} = 0.4587° - 0.5290° = -0.0703°$$
$$= -4.21 MOA.$$

Change in angle of sight-scope is
$$\Delta\alpha_h = -\Delta\alpha_S = -(-4.21) = 4.21 MOA.$$

The number of sight clicks is
$$\Delta N = 0.421 / 0.125 = 33.74 \approx 34 clicks.$$

We have to reduce the sight scope clicks by 34 clicks.

2.2 Basic Formula for Inclined Shooting

The above approximate formula (2.1.1) is derived from the following formula[6]
$$\sin(2\alpha_{OI} + E) = \sin(2\alpha_0) \cdot \cos^2(E) + \sin(E). \qquad (2.2.1)$$

Formula (2.2.1) is determined using the trajectory equation of the projectile flying in absence of drag[7].

Hence, for the super elevation angle we can write:
$$\alpha_{OI} = \frac{\arcsin[\sin(2\alpha_0) \cdot \cos^2(E) + \sin(E)] - E}{2}. \qquad (2.2.2)$$

Note

As it is shown in the example 2.2, using the formula (2.2.2) we can see that the super declined angle is slightly smaller than the super inclined angle. That discrepancy is result of the term, which is positive for the uphill shooting and negative for the downhill shooting. That result is normal since in uphill shooting the gravity acts "against" the "climbing" bullet while in declined shooting the gravity acts in "favor" of "descending bullet"

It can easily be seen that equation (2.1.1) calculates the same value for the super elevation angle and super depression angle.

Example 2.1

Find the super elevation angle for the given Lapua GB528 19.44 bullet, if the inclined range is 1,000 yards and the inclined angle is 30 degree.

Solution

Substituting in (2.2.2) $\alpha_0 = 0.5247°$, $E = 30°$, we find that the super elevation angle is:

$$\alpha_{OI} = \frac{\arcsin[\sin(2 \cdot 0.5247) \cdot \cos^2(30) + \sin(30)] - 30}{2} = 0.4565°.$$

Using the approximate formula 1 we have found that the super elevation angle (table 1.1) is $\alpha_{OI} = 0.4544°$. Between the estimated elevation angles, calculated respectively using formula (2.1.1) and formula (2.2.2), there is an insignificant difference of

$$\Delta\alpha_{OI} = 0.4565 - 0.4544 = 0.0021° = 0.126 MOA,$$

or approximately 1 click.

Using formula (2.2.2), in the following table 2.1, there is estimated the departure angle for Lapua Scenar GB528, 19.44g bullet, fired in ICAO atmosphere with velocity 830m/s = 2723.10fps.

Table 2.1 Super Elevation Angle, Lapua Scenar GB528, 19.44g, ICAO Atmosphere.

Range (yards)	Elevation Angle (Degree)											
	0	10	15	20	25	30	35	40	45	50	55	60
	Super Elevation Angle (Degree)											
100	0.0379	0.0366	0.0356	0.0356	0.0344	0.0328	0.0311	0.0290	0.0268	0.0244	0.0218	0.0190
200	0.0789	0.0762	0.0742	0.0742	0.0715	0.0684	0.0647	0.0605	0.0558	0.0508	0.0453	0.0395
300	0.1218	0.1177	0.1145	0.1145	0.1106	0.1056	0.0999	0.0934	0.0863	0.0784	0.0700	0.0610
400	0.1688	0.1632	0.1588	0.1588	0.1532	0.1464	0.1385	0.1296	0.1196	0.1087	0.0971	0.0846
500	0.2178	0.2106	0.2049	0.2049	0.1977	0.1890	0.1788	0.1673	0.1544	0.1404	0.1253	0.1093
600	0.2709	0.2620	0.2550	0.2550	0.2460	0.2352	0.2225	0.2082	0.1922	0.1748	0.1560	0.1360
700	0.3273	0.3166	0.3082	0.3082	0.2974	0.2843	0.2690	0.2517	0.2324	0.2113	0.1886	0.1645
800	0.3877	0.3751	0.3652	0.3652	0.3524	0.3369	0.3188	0.2983	0.2755	0.2505	0.2236	0.1950
900	0.4536	0.4389	0.4273	0.4273	0.4124	0.3943	0.3732	0.3492	0.3225	0.2933	0.2618	0.2283
1000	0.5247	0.5080	0.4946	0.4946	0.4774	0.4565	0.4321	0.4043	0.3734	0.3397	0.3032	0.2645
1100	0.6005	0.5816	0.5663	0.5663	0.5467	0.5228	0.4949	0.4631	0.4278	0.3891	0.3474	0.3030
1200	0.6831	0.6619	0.6445	0.6445	0.6222	0.5951	0.5634	0.5273	0.4871	0.4431	0.3957	0.3451
1300	0.7731	0.7493	0.7298	0.7298	0.7047	0.6741	0.6382	0.5974	0.5519	0.5021	0.4484	0.3911
1400	0.8698	0.8435	0.8216	0.8216	0.7934	0.7590	0.7188	0.6729	0.6217	0.5657	0.5052	0.4407
1500	0.9758	0.9467	0.9223	0.9223	0.8908	0.8524	0.8072	0.7558	0.6984	0.6356	0.5677	0.4952
1600	1.0913	1.0593	1.0322	1.0322	0.9971	0.9542	0.9039	0.8464	0.7823	0.7119	0.6359	0.5549

Note

The method actually in use by some ballisticians, or shooters to predict the trajectory of bullets in uphill or downhill shooting, involves the estimation of the "Equivalent Horizontal Range" (EHR). The EHR method is the basis to compensate for ballistic drop of the bullet[8].

The approximate method based on EHR is usually called the "rifleman's rule". Rifleman's rule is obtained for the projectile flying in the uniform gravitational force field and in absence of air resistance (see end note, ref. 4).

The rifleman's rule is not so easy to be applied in practice of shooting, especially for long ranges. Moreover, for long ranges, the accuracy of estimation of the sighting angle is not satisfactory.

Downhill Shooting

The depression angle E for downhill shooting is negative i.e. $E < 0$.

In the same way as in the inclined shooting, using formulas (2.1.1) or (2.2.2), we can obtain similar tables as in the inclined shooting.

As we will see in the following example, the declined angle of shooting is almost the same as the super-elevation angle in inclined shooting.

Example 2.2 Lapua GB528 19.44 bullet

For the declined range $R_S = 1000$ yards, and the depression angle $E = -30$ degree, using (2.2.2) we find that the declined angle of shooting is

$$\alpha_{OI} = \frac{\arcsin[\sin(2 \cdot 0.5247) \cdot \cos^2(-30) + \sin(-30)] - (-30)}{2}$$

$$= 0.4523°.$$

Comparing the above value $\alpha_{OI} = 0.4523°$ for the declined shooting with the corresponding value of inclined shooting $\alpha_{OI} = 0.4564°$ (given in table 2.1), we see that there is a difference, though relatively small.

2.3 Drop, Super Elevation Angle and Inclined Range

Okunev's Approach

According to Okunev, we assume that at the inclined range,

$$R = x_T \cdot \frac{\cos(E + \alpha_{OI})}{\cos(\alpha_{0T}) \cdot \cos(E)}, \qquad (2.3.1)$$

the inclined drop of a bullet, $Drop(R_1 R)$, is equal to the bullet horizontal drop at $x_T = OT$,

$$\bar{y}_T = Drop(T_1 T)\,^9,$$

i.e.

$$drop(RR_1) \approx \bar{y}_T. \qquad (2.3.2)$$

while the super elevation angle is estimated by the equation:

$$\alpha_{OI} = \alpha_{0T} \cdot \cos(E), \qquad (2.3.3)$$

where $R = 0R$, $x_T = 0T$ and elevation angle is E.

Formulae (2.3.1) - (2.3.3) are valid also for downhill shooting, with depression angle E negative.

Note that formulas (2.3.1) - (2.3.3) are obtained for the projectile flying in presence of drag, on conditions that the initial velocity v_0 of the projectile and the ballistics coefficient BC are constant.

It is known that BC of a projectile that corresponds to a well known G-function (G_1, G_7, etc.) does depend on the departure angle.

The BC related with a characteristic G-function (for example the personalized G-function of GB528 Lapua Scenar bullet considered in our examples) is constant and does not depend on the departure angle $(\alpha_{OI} + E)$.

Since many standard range tables are constructed using G_1 or G_7 functions of resistance and fixed BC, the formulas (2.3.1) - (2.3.3) are valid as well for the standard range tables calculated using G_1, or G_7 functions of resistance.

The formula of (2.3.3) is identical to formula (2.1.1), but obtained in a different way (see Okunev, B. H "Fundamentals of Ballistics, p. 240, Vol. 1, Book 2, Moscow, 1943).

Formulae (2.3.1) - (2.3.3) show that when we are shooting with super elevation angle α_{OI}, the inclined shooting range is somewhat different from the corresponding horizontal range.

As we will demonstrate in the following examples, for uphill shooting the inclined range $OR < x_T$, while for downhill shooting the downhill range $OR > x_T$.

In other words, in uphill shooting, if the target is at an inclined range that is equal to horizontal range x_T, the projectile will hit the target before reaching the center, while in downhill shooting the projectile will exceed the distance from the muzzle to the center of the target (the impact point is above the center of the target).

Let's illustrate the use of equations (2.3.1) - (2.3.3).

Example 3.1

A Lapua GB528 19.44 bullet is fired uphill/downhill with departure velocity 830m/s.

(a) The inclination angle is E = 30 degree and the firing uphill distance is 1,500 yards.

Find the inclined shooting angle and the point of impact (uphill distance of the point of impact).

For the horizontal range 1,500 yards, $\alpha_{0T} = 0.9758°$.

(b) The depression angle is E = - 30 degree and the firing downhill distance is 1,500 yards.

Find the declined shooting angle and the point of impact (downhill distance of the point of impact on the declined plane).

For the Lapua bullet in study, the ballistic coefficient relative to the Lapua G-function of resistance is constant, equal to $c = 3.796 m^2 / kg$.

Solution

(a) Substituting in (2.3.3) we find that the elevation angle is

$$\alpha_{OI} = \alpha_{OT} \cdot \cos(E) = 0.9758 \cdot \cos(30) = 0.8451°.$$

Shooting uphill with the calculated angle, the inclined range will be:
$$R = x_T \cdot \frac{\cos(E + \alpha_{OI})}{\cos(E) \cdot \cos(\alpha_0)} = 1500 \cdot \frac{\cos(30 + 0.8451)}{\cos(30) \cdot \cos(0.9758)}$$

$$= 1487.28 \, yard.$$

The bullet will hit the ground $\Delta R = 1487.28 - 1500 = -12.72$ yards before reaching the center of target.

The angle of the bullet impact on horizontal shooting at range x_T is always somewhat greater than the corresponding departure angle α_{OT}.

The same statement holds for impact angle in inclined shooting.

So, if we consider the impact angle equal to the super elevation angle, i.e. $\alpha_{OI} = 0.8451°$, then the bullet will hit

$$\Delta y = -12.72 \cdot \tan(0.8451) = -0.188 \, yd. = -6.77 \, in.$$

below the center of the target located at 1500 yards.

We need to correct the super elevation angle by adding an insignificant correction

$$\Delta \alpha_{OI} = \arctan(0.188/1500) = 0.007° = 0.433 MOA.$$

The corrected inclined angle will be
$$\alpha_{OI}(new) = \alpha_{OI} + \Delta \alpha_{OI} = 0.8451 + 0.007° = 0.852°.$$

(b) Employing (2.3.1) for the downhill shooting we have:
$$\alpha_{OI} = \alpha_{OT} \cdot \cos(E) = 0.9758 \cdot \cos(-30) = 0.8451°$$

Shooting downhill (E = -30 degree) with the calculated angle $\alpha_{OI} = 0.8451°$, we find that the declined range will be

$$R_S = R_T \cdot \frac{\cos(E + \alpha_{in})}{\cos(E) \cdot \cos(\alpha_0)} = 1500 \cdot \frac{\cos(-30 + 0.8451)}{\cos(-30) \cdot \cos(0.9758)}$$

$$= 1512.83 \, yards.$$

The bullet in downhill shooting will impact
$$\Delta R_S = 1512.83 - 1500 = +12.83 \, yards$$

behind the center of target.

If we consider the impact angle is equal to the super elevation angle, i.e. $\alpha_{OI} = 0.8451°$, then the bullet will hit the target
$$\Delta y = 12.83 \cdot \tan(0.8451) = 0.189 \, yd. = 6.84 \, in$$

above the center of the target.

We need to correct the elevation angle by subtracting an insignificant correction value
$$\Delta \alpha_{OI} = \arctan(0.189/1500) = 0.007° = 0.433 \, MOA.$$

The corrected declined angle is
$$\alpha_{OI}(new) = \alpha_{in} - \Delta \alpha_{OI} = 0.8451 - 0.007° = 0.838°.$$

Comment

The deviation of the projectile down (-6.77 inches) or above (+6.84 inches) the center of the target is insignificant considering the long range 1500 yards of the shooting.

The inclined angle correction of 0.433 MOA is insignificant as well.

2.4 Rigidity Principle of Projectile Trajectory

For relatively small elevation or depression angles (E), we can use the principle of rigidity of the trajectory to estimate the inclined or declined angle of shooting.

Rigidity Principle

The trajectory of a projectile can be rigidly rotated up or down around the muzzle without significantly changing its form, provided that "the elevation or depression angle is not greater than 20 meters in vertical"[10], (or 65.62 ft = 21.872 yards).

Thus, the "rigid" elevation or depression angle must not be greater than the angle E_R determined by the relation

$$\tan E_R = \pm 20 / R_T. \tag{2.4.1}$$

In imperial units (R_T in yard):

$$\tan E_R = \pm 21.872 / R_T. \tag{2.4.2}$$

(The "+" sign belong to elevation angle (uphill shooting), while the "-" sign belong to downhill shooting.)

Thus, for small elevation/depression angle E, the form of the trajectory and all related characteristics do not change significantly with inclined/declined angle E if $|E| \leq |E_R|$.

For relatively small elevation/depression angles (determined by (2.4.1), or by (2.4.2)) the super elevation angle is equal to the horizontal departure angle, i.e. $\alpha_{in} = \alpha_0$.

Example 4.1

For the downhill ranges 100 yards, 400 yards, 800 yards and 1500 yards find the corresponding declination angle E_R that corresponds to the maximum rigid rotation of the horizontal trajectory.

Solution

Equation (2.4.2), for the downhill shooting, can be written:

$$E_R = -\arctan(21.872 / R_T). \tag{2.4.3}$$

Substituting, 100 yards, 400 yards, 800 yards and 1500 yards in (8) we find respectively:

$$E_R = -\arctan(21.872/100) = -12.3374°,$$

$$E_R = -\arctan(21.872/400) = -3.1298°$$

$$E_R = -\arctan(21.872/800) = -1.5661°,$$

$$E_R = -\arctan(21.872/1500) = -0.8354°.$$

So, for the inclined or declined range 100 yards and for the elevation or depression angle in absolute value less than or equal to the absolute value of $E_R = \pm 12.3374°$ we can use the same angle of sight that we set up for the same 100 yards horizontal shooting.

For the inclined or declined long range shooting at 1500 yards we can use the same angle of sight-scope set up for horizontal range 1500 yards, if the elevation or depression angle is in absolute value less than the absolute value of $E_R = -0.8354°$.

Practically, for shooting even for relatively moderate inclined ranges, the rigidity principle can not be applied.

2.5 Non-Rigidity Principle of Trajectory

Equation (2.1.1) estimates the super elevation angle on condition that the inclined range is equal to the corresponding horizontal range, i.e.

$$\alpha_{OI} = \alpha_{0T} \cdot \cos(E), \quad (2.5.1)$$

considering that

$$OR = x_T.$$

Projectile Drop

Let's calculate the projectile drop $\overline{y}_{R1} = RR_1$ in inclined shooting (fig. 5, section 2.1). The angle $\angle R_2 R R_1$ is equal to the elevation angle E, i.e. $E = \angle R_2 R R_1$.

Referring to the right triangle $R_2 R R_1$, we have:

$$|\overline{y}(RR_1)| = \frac{RR_2}{\cos E}. \quad (2.5.2)$$

Referring to triangle ROR_2, we can write:

$$RR_2 = OR \cdot \tan(\alpha_{OI}) \approx x_T \cdot \alpha_{OI} \cdot \frac{\pi}{180} = x_T \cdot \alpha_{OT} \cdot \cos(E) \cdot \frac{\pi}{180}. \qquad (2.5.3)$$

Substituting (2.5.3) into (2.5.2), we find that inclined drop is equal to the horizontal drop $\bar{y}_T = TT_1$, i.e.

$$\bar{y}(RR_1) \approx x_T \cdot \alpha_{OT} \cdot \frac{\pi}{180} = \bar{y}_T. \qquad (2.5.4)$$

when

$$OR = x_T \text{ and } \alpha_{OI} = \alpha_{OT} \cdot \cos(E). \qquad (2.5.5)$$

Apparent Drop

In inclined shooting (fig. 5, section 2.1), the **apparent drop** ($ADrop = \bar{y}'(RR_2)$) is the vertical distance RR_2, i.e.

$$\bar{y}'(RR_2) = \bar{y}(RR_1) \cdot \cos(E). \qquad (2.5.6)$$

which, since $\bar{y}(RR_1) = \bar{y}_T$, in absolute value can be written

$$|\bar{y}'(RR_2)| = |\bar{y}_T| \cdot \cos(E). \qquad (2.5.7)$$

Thus, while the projectile drop in inclined shooting remains equal to the corresponding horizontal drop the absolute value of the apparent drop decreases with the elevation angle E according to equation (2.5.6).

The uphill or downhill shooting, considered in sections (2.4) is related to the principle of rigidity of trajectory.

For large inclination or declination angles that do not satisfy the condition of trajectory rigidity, $|E| \leq |E_R|$, i.e. when the elevation or depression angle in absolute value is not smaller than the absolute value of E_R determined by (2.4.1), (2.4.2) or (2.4.3), the principle of trajectory rigidity is not valid.

In these cases we can use the equation (2.5.1) to estimate the super elevation angle. As a matter of fact the form of the trajectory continuously changes, but the vertical drop of projectile does not change. Moreover the apparent drop changes according to equation (2.5.6).

We can formulate the following statement that we call Non-Rigidity of Trajectory Principle:

Non-Rigidity Principle

If we rotate the horizontal range $x_T = OT$ at an angle equal to elevation angle E, the form of projectile trajectory changes in such a way that the **inclined drop** of projectile remains equal to the **horizontal drop**, but the **apparent drop** decreases proportionally with decreasing of $\cos(E)$, i.e. in accordance with equation (2.5.6).

Based on the Non-Rigidity Principle we can estimate the apparent drop and the super elevation angle using respectively the equation (2.5.7) and the equation:

$$\alpha_{OI} \approx \frac{|\bar{y}'(RR_2)|}{x_T} \cdot \frac{180}{\pi}. \qquad (2.5.8)$$

The last equation is equivalent to the compact equation (2.5.1), i.e.

$$\alpha_{OI} = \alpha_{0T} \cdot \cos(E).$$

Formulae (2.5.6) and (2.5.8) express the **Non-Rigidity Principle and Equal Drop Model** that allows us to find super elevation or depression angle in uphill or downhill shooting if we know the drop of projectile in horizontal shooting, or equivalently if we know the departure angle α_{0T} in horizontal shooting.

Example 5.1

A bullet is fired to hit a target located 1200 yards from the rifleman, at an inclined range 25 degree.

Use the Non-rigidity principle to find the apparent drop and the super elevation angle if the drop of the projectile at horizontal range 1,200 yards is 14.31 yards.

Solution

Apparent drop is

$$|\bar{y}'(RR_2)| = |\bar{y}_T| \cdot \cos(E) = |-14.31| \cdot \cos(25) = 12.97 \text{ yards.}$$

The super elevation angle is

$$\alpha_{OI} = \frac{|\bar{y}'(RR_2)|}{x_T} \cdot \frac{180}{\pi} = \frac{12.97}{1200} \cdot \frac{180}{\pi} = 0.6191°$$

2.6 Shooting with Departure Angle Zero

Let's suppose that the bullet is fired horizontally with departure angle $\alpha_O = 0°$.

The bullet trajectory is under the x-axis. At the horizontal range $x_T = OT$ the bullet drop has a certain value $\bar{y}_{OT} = T_0 T$.

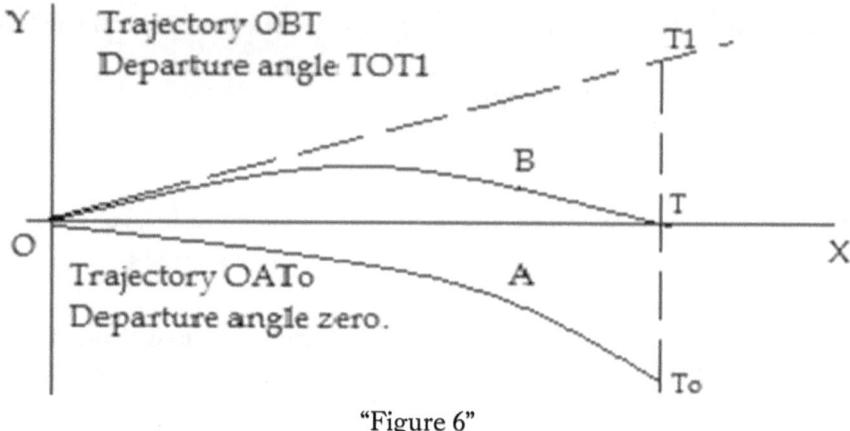

"Figure 6"

Suppose now that $\alpha_{0T} = \angle TOT_1$ is the departure angle that zeroes the firearm at the horizontal range $x_T = OT$. The bullet drop is $\bar{y}_T = TT_1$.

According to the Non-Rigidity of Trajectory model, the bullet drop is the same for both trajectories, i.e.

$$\bar{y}_T = \bar{y}_{OT}. \tag{2.6.1}$$

So, for the departure angel angle α_{0T}, referring to the right triangle TOT_1 we can write:

$$\tan(\alpha_{0T}) = \frac{|\bar{y}_{OT}|}{x_T}, \tag{2.6.2}$$

or approximately

$$\alpha_{0T} \approx \frac{|\bar{y}_T|}{x_T} \cdot \frac{180}{\pi}. \tag{2.6.3}$$

since both drops are equal (equation (2.6.1)).

… ELEMENTS OF EXTERIOR BALLISTICS

Thus:

- If at a certain horizontal range $x_T = OT$, we know the bullet drop $\bar{y}_{OT} = T_0 T$ when the bullet is fired with departure angle zero $\alpha_O = 0°$, then the angle of departure α_{0T} needed to zero the firearm at the given horizontal range $x_T = OT$ is predicted using equation (2.6.3).

The departure angle in MOA (Minute of Angle) is

$$\alpha_{0T} \approx \frac{|\bar{y}_T|}{x_T} \cdot \frac{10800}{\pi}. \tag{2.6.4}$$

Comment

In some range tables is not given the departure angle but only the bullet drop that corresponds to a given horizontal range. The departure angle is unknown.

In such cases we can assume that the bullet is fired horizontally (departure angle zero, or that the bullet is fired at whatever super elevation angle.

For example, during radar tests the bullet is fired at a given elevation E with a given super elevation angle, that might be known to us or not.

Using the Non-rigidity trajectory model, we can find the departure angle that zeroes the firearm at a given horizontal range employing equation (2.6.3).

Example 6.1

The 0.338 Lapua GB528 bullet is fired horizontally. The projectile drop at the horizontal range 1,500 meters is - 30.035 meters[11].

(a) Find the departure angle α_{0T} that zeroes the firearm at 1500 meters.

(b) Find the corresponding angle of sight, if the sight height is $h = 2.7" = 2.7 \cdot 0.0254 = 0.06858 m$.

(c) For the same inclined range (1,500m), find the super elevation angle if the elevation angle is E = 30 degree as well as the corresponding aiming angle.

(d) Find the super depression angle if E = - 30 degree.

Solution

Since the departure angle is not given at the Wikipedia paper, we can assume that the bullet is fired horizontally (departure angle zero), or, we can assume that the bullet is fired at whatever super elevation angle.

(a) Substituting in (2.6.3) we find that the departure angle that zeroes the firearm at horizontal range 1500 meters is

$$\alpha_{OT} \approx \frac{|y_T|}{x_T} \cdot \frac{180}{\pi} = \frac{30.035}{1500} \cdot \frac{180}{\pi} = 1.1473°.$$

(b) Substituting in (2.1.4), we find that the aiming angle is

$$\alpha_S = \alpha_{OT} + \frac{h}{R_S} \cdot \frac{180}{\pi} = 1.1473° + \frac{0.06858}{1500} \cdot \frac{180}{\pi} = 1.150°$$

$$= 68.99 MOA.$$

(c) Employing (2.1.1), we find that the super elevation angle is
$$\alpha_{OI} = \alpha_{OT} \cdot \cos(E) = 1.1473 \cdot \cos(30) = 0.9936°.$$

(d) The super depression angle is
$$\alpha_{OI} = \alpha_{OT} \cdot \cos(E) = 1.1473 \cdot \cos(-30) = 0.9936°.$$

2.7. Numerical and Approximate Solutions

The super elevation or depression angle and the other elements of the projectile trajectory can be obtained solving the differential equations of point-mass projectile trajectory[12]:

$$\begin{cases} \dfrac{dv_x}{dx} = -c \cdot \dfrac{p_0}{p_{oN}} \cdot \sqrt{\dfrac{\tau_{0N}}{\tau_0}} \cdot h(y) \cdot \dfrac{G_D(v)}{v} \\ \dfrac{dp}{dx} = -\dfrac{g}{v_x^2} \\ \dfrac{dt}{dx} = \dfrac{1}{v_x} \\ \dfrac{dy}{dx} = p \end{cases} \quad (2.7.1)$$

where:

α is the angle (in degree) the projectile velocity forms with x-axis, ($p = \tan \alpha$), (x, y) are the coordinates of projectile at a moment t, v is the projectile velocity, c is the ballistic coefficient (BC), $G_D(v)$ is the function of resistance of the given projectile and $g = 9.0665 m/s^2$ is the gravity constant.

The density function $h(y)$ at the sea level is 1, i.e. $h(y) = 1$.

The component of velocity along x-axis is $v_x = v \cdot \cos \alpha$.

For the 0.338 **Lapua Scenar GB528 19.44 g (300 gr.)** bullet launched at an angle $(\alpha_{in} + E)$ with velocity $v_0 = 830 m/s$ in ICAO atmosphere, using methods of numerical integration (5) we find the super elevation angle α_{in}.

The ballistic coefficient of the given Lapua bullet and the characteristic drag function[13] (in ICAO atmosphere) are respectively $c = 3.796 m^2/kg$ and

$$G_D(v) = 0.141 \cdot v - 30.039, \qquad (2.7.2)$$

for $325 < v < 850$,

$$G_D(v) = 0.02438 \cdot v - 1.3696, \qquad (2.7.3)$$

for $v \leq 325$

The characteristic G-function of 0.338 GB528 Lapua Scenar in ASM atmosphere is

$$G_D(v) = 0.139 \cdot v - 29.709, \qquad (2.7.4)$$

for $330 < v < 850$

$$G_D(v) = 0.0242 \cdot v - 1.399, \qquad (2.7.5)$$

for $v \leq 330$

Note

Since the G-function of resistance (2.7.2) is found using Doppler radar data, the horizontal range obtained using the system of differential equations (2.7.1) is accurate. There is no need to introduce range corrections for the rotation of the projectile.

The drift to the right, or to the left of firing plane, due to the rotation of the projectile, must be calculated.

Accuracy of the Non-Rigidity- Principle and Equal Drop Model

In table 7.1 there are given the Doppler radar measurements for Lapua Scenar GB528 19.44 g (300 gr.) bullet (end note 2).

Table 7.1. Lapua Scenar GB528 19.44 g (300 gr.) bullet (Wikipedia)

Range (m)	0	300	600	900	1,200	1,500
Velocity (m/s)	830	711	604	507	422	349
Time (s)	0	0.3918	0.8507	1.3937	2.0435	2.8276
Bullet Drop (m)	0	0.715	3.203	8.146	16.571	30.035

Using the drop, given in Table 7.1, and the Non-Rigidity-Principle (formula 2.5.8), we have estimated the inclined super elevation angle for elevation angle 5, 10, 15, 20, 30, 40 degree.

In table 7.2, it is added the super elevation angle for the range 100 meters, estimated using the drop 0.074m shown in last column[14].

Table 7.2. Lapua GB528 19.44g bullet. Non-Rigidity-Principle

Range	100	300	600	900	1200	1500
Elevation	Super Elevation Angle [degree]					
0	0.0424	0.1366	0.3059	0.5186	0.7911	1.1471
5	0.0422	0.1360	0.3046	0.5162	0.7872	1.1407
10	0.0418	0.1344	0.3009	0.5099	0.7773	1.1258
20	0.0398	0.1282	0.2869	0.4858	0.7400	1.0706
30	0.0367	0.1181	0.2642	0.4971	0.6805	0.9836
40	0.0325	0.1044	0.2335	0.3950	0.6007	0.8676
Drop	0.074	0.715	3.203	8.146	16.571	30.035

In the following table 7.3 there is given the super elevation angle obtained by solving numerically the system of equations (2.7.1).

Table 7.3. Numerical Integration. Lapua GB528 19.44.

| Range | 300 | 600 | 900 | 1200 | 1500 |

Elevation	Super Elevation Angle [degree]				
0	0.1370	0.3061	0.5139	0.7920	1.1485
5	0.1360	0.3057	0.5117	0.7890	1.1420
10	0.1350	0.3015	0.5110	0.7764	1.1280
20	0.1285	0.2875	0.4876	0.7442	1.0684
30	0.1190	0.2700	0.4485	0.6825	0.9922
40	0.0328	0.2355	0.3972	0.6018	0.8750
Drop	0.715	3.203	8.146	16.571	30.035

Comparing super elevation angle in table 7.2 and table 7.3, we can see that "Non-Rigidity-Principle" of projectile trajectory gives satisfying and accurate trajectory prediction.

Knowing the scope height, we can find the angle of sight-scope in MOA, for the uphill or downhill shooting and then the numbers of sight clicks needed to adjust shooting.

We recommend to the shooter to prepare a card for the super elevation or depression angle based on the shooting data for the horizontal range and then using the Non-Rigidity-Drop model (Formulae (2.5.6) or (2.5.8)).

In table 7.4, there is given the super-elevation angle for ranges in Imperial units (yard) obtain using formula (2.5.1).

As an input to (2.5.1) we use the horizontal departure angle, for elevation or depression angle $E = 0$.

Table 7.4. Lapua GB528 19.44g bullet. Non-Rigidity. Imperial Units (yard).

Range	300	600	900	1,200	1,500
Elevation		Super Elevation Angle			
0	0.1218	0.2709	0.4535	0.6831	0.9758
5	0.1213	0.2698	0.4515	0.6798	0.9706
10	0.1199	0.2666	0.4460	0.6713	0.9581
20	0.1144	0.2542	0.4250	0.6393	0.9117
30	0.1055	0.2341	0.3912	0.5881	0.3880
40	0.0932	0.2069	0.3456	0.5193	0.7394

In general, comparing the data in table 7.1 and table 7.4, we can see that the approximate formula 2.1.1 (as well as formula (2.1.4) and equal-drop and non-rigidity model) can be always used for approximate calculations.

Formula (2.1.1) is convenient for its practical simplicity.

Summary

We have introduced some practical simple models that allow the shooters to set up the rifle to shoot uphill or downhill.

Any long range shooter, for training purposes, can prepare his or her own elevation shooting card based on the method presented in this short paper.

The shooter can as well prepare an app to find automatically the angle of sight in inclined or declined shooting.

The long range marksman can use his own range tables (for $E = 0$) that are obtained in the practice of field shootings, and based on that table the marksman can prepare the inclined or declined shooting range table.

2.8 Exterior Ballistics PC Programs

There are three main Exterior Ballistics PC programs associated with the book. The PC programs (codes), in quick basic (QB), were compiled in 1991 by Col. Genc Kokoshi (ex-professor at the Academy of General Staff, Tirana). The PC programs are continuously modified by the author, to reflect the advancements in Exterior Ballistics.

The three initial PC programs were based on Siacci's G-function and were valid only for a type of artillery projectile, when field cannon shooting is done in TSA atmosphere.

The core of all PC programs developed since 2005 are the code's that Col. Kokoshi has designed in 1991.

The QB codes designed by Col. Kokoshi are fascinating compact codes that made possible for me to advance in the study of exterior ballistics.

The Programmable TI58c, I was using in 80's, had no capacity to solve complex exterior ballistics problems though it was an amazing technology that helped me in my studies, solving few exterior ballistics problems related with the trajectory of projectile launched with low velocities, under 250 m/s.

The PC programs are compiled based on the system of differential equations (2.7.1). There are three types of PC programs.

1. The PC program that calculates the BC of a given projectile with respect to reference G_1, G_7 or other reference G-function of resistance.

The PC program named BC2016.Bas is shown in Appendix C.

2. The PC program that calculates the departure angle needed to hit a given target located at a given point.

   The PC program, named Alapua16.BAS, is shown in appendix D.

   The program calculates the departure angle needed to hit a given target.

   The program uses the characteristic G-function of 0.338 GB528 19.44, 8.59 mm, velocity 830m/s, BC = 3.796 (Appendixes B, B1 and B2).

   The PC program APROJ16.BAS, also shown in appendix D, calculates the departure angle using the main reference G-functions, G_1, G_2, ...G_7, Siacci's and Russian year 1943, G_{43}-function (Appendix A).

3. The PC program ACHA16.BAS, shown in appendix D, is used to find the departure angle and other elements of the trajectory for the following list of bullets (Appendix B).

List of Bullets

1. Bullet 0.300 Winchester Magnum, Velocity 884m/s, BC = 4.97096:

2. Caliber 0.308, 168 grain, Sierra International Bullet, Velocity 792.48m/s, BC = 5.6325.

3. M118LR Bullet, 0.308" mass 0.01134 kg. Velocity 884m/s, BC = 5.3970.

4. M118 Ball Bullet (Federal GM308M2), 0.308", mass = 0.00134kg, velocity 792.48m/s, BC = 5.3973

5. 300 gr., .338 - .416 Bullet (mass 0.01944kg, Velocity 927.40m/s, BC = 3.791

6. Caliber 0.30 Ball M2 Bullet, mass 0.00972 kg, Velocity 853.44 m/s, BC = 6.2965

For each PC program there are given examples to demonstrate the use of the program.

4. There are three PC programs that estimate the projectile range given the departure angle. They are named RPROJ16.BAS, RLAPUA16.BAS and RCHA16 shown in Appendix E.

The first program predicts the elements of trajectory using the main reference G-functions, G_1, G_2, ...G_7, Siacci's and Russian year 1943, G_{43}- function.

RLAPUA16.BAS uses the characteristic G-function of GB528 Lapua Scenar bullet (Appendix B, G-functions B1 and B2).

The PC program RCHA16 predicts the point of impact when it is known the departure angle. It is valid for the list of bullets, presented in appendix B, and listed above in 3.

All three programs predict as well the projectile drop in any point when launching angle is zero.

Range Wind and Cross Wind Included in the Differential Equations

To include the wind effect in differential equations of projectile trajectory we can use the McCoy's method shown in his book "Modern Exterior Ballistics".

In preparing the PC programs in our book we use the method based on Didion's equations on the relative velocity of projectiles.

Range wind effect

We follow the Didion's method that considers the system of coordinates (xoy) moving with the velocity of range-wind $\vec{w}_x$. In that system of coordinates, the resistance of air does not depend on the velocity of range wind, but it depends on the velocity of the projectile with respect to the atmospheric air.

So, the characteristic G-function of resistance in the moving system of coordinates (xoy) is the same.

Now, we consider a system of coordinates (xoy) moving with the velocity of wind, and a fixed system of coordinates (X0Y). At the initial moment of shooting both systems have the origin, at the muzzle of firearm.

To find the trajectory of projectile in presence of range-wind (fig. 7), in the differential system of equations (2.7.1) we have to consider that in the moving system of coordinates (xoy), the projectile is launched with the relative initial velocity

$$v_{0r} = \sqrt{(v_0^2 - 2v_0 w_x \cdot \cos\alpha_0 + w_x^2)}, \qquad (2.8.1)$$

and with launching angle

$$\alpha_{0r} = \arctan\left(\frac{v_0 \sin \alpha_0}{v_0 \cos \alpha_0 - w_x}\right).\qquad(2.8.2)$$

When the wind is blowing in the positive direction of x-axis, the velocity of wind is positive ($w_x > 0$). Otherwise, the wind velocity is negative, $w_x < 0$.

Horizontal shooting

After solving the differential equations (2.7.1) with initial conditions (2.8.1) and (2.8.2), to find the impact coordinate of the projectile with respect to the coordinate system related with the ground (X0Y), we need to add to the range x_T in the moving system of coordinates the quantity $w_x \cdot t_T$, i.e. the impact range is

$$X_T = x_T + w_x \cdot t_T.\qquad(2.8.3)$$

where x_T, t_T are respectively the impact range of projectile on the ground, and time of impact in the moving system of coordinates while X_T is the impact range with respect to the fixed system. with respect to m

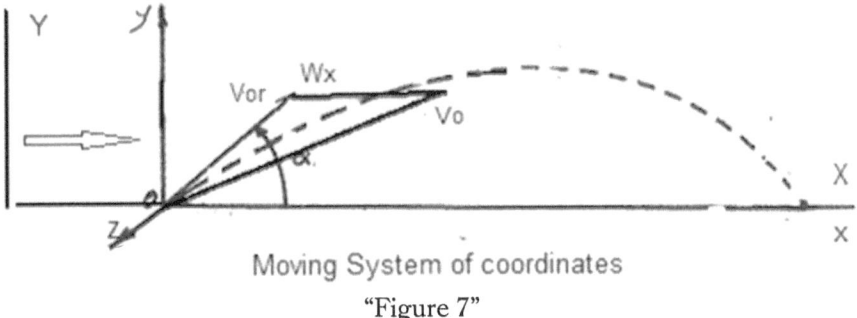

Moving System of coordinates

"Figure 7"

Uphill shooting

In inclined uphill shooting, to find the coordinates of the impact point, we need to employ at the end of solution of (2.7.1), the equation (2.8.3) and the equation

$$Y_T = y_T + \tan \alpha_0 \cdot w_x \cdot t_T.\qquad(2.8.4)$$

where y_T and t_T are respectively the x-coordinate and the time of flight of projectile (in the moving system of coordinates) till at the inclined point of impact.

The following examples illustrate the solution of exterior ballistics problems using PC programs.

There are detailed instructions, illustrated with examples, on how to use all the PC programs at the beginning of each program code.

Declined and Horizontal Shooting

The PC programs, Alapua16.bas, APROJ16.Bas, and ACHA16.Bas can predict the declined angle of shooting, using PC programs. Using the other three PC programs, RPROJ16.BAS, RLAPUA16.BAS, and RCHA16 we can estimate the elements of trajectory in inclined or declined shooting

Example 8.1

(a) Find the ballistics coefficient c_7 related to the reference G_7-function for the 0.338 GB528 Lapua Scenar bullet launched in ICAO standard atmosphere with velocity $v_0 = 830 m/s$ if the point of impact is at the horizontal range $x_T = 1500 m$ from the muzzle. There is no wind. The departure angle is 1.1471 degree.

(b) Find as well the BC for ranges 1200m, 900m, 600 m if the departure angle is respectively 0.7912 degree, 0.8186 degree, 0.3059 degree

(c) For the given bullet find the ballistics coefficient c_1 related to the reference G_1-function for the horizontal range $x_T = 1500 m$.

Solution

(a) Using BC2016.Bas (see the instructions in the PC program) we find:
$c_7 = 3.6842 m^2 kg$.

We find as well impact velocity 352.97m/s, time of flight 2.8258s.

In Imperial units BC is

$$C_7 = \frac{1.4222}{c_7} = \frac{1.4222}{3.6842} = 0.3860 lb/in^2.$$

(b) Running the program with the given zero range and the respective departure angle we find respectively:

Range 1200m: $c_7 = 3.690 m^2 kg$, impact velocity 420m/s,

Range 900m: $c_7 = 3.721 m^2 kg$, impact velocity 502m/s.

Range 600m: $c_7 = 3.789 m^2 kg$, impact velocity 597m/s.

(c) The BC related with G_1-function is: $c_1 = 1.8683 m^2 kg$.

Example 8.2

Find the ballistics coefficient c_7 related to the reference G_7-function for the 0.338 GB528 Lapua Scenar bullet launched in ICAO standard atmosphere with velocity $v_0 = 830 m/s$ if the drop of the bullet at the horizontal range $x_T = 1200 m$ is $\bar{y} = -16.571 m$.

Solution

Use BC2016.Bas. Input the departure angle zero.

Note that when departure angle is zero, we have to input the y-coordinate of the muzzle of the firearm $\bar{y} = +16.571 m$ over the ground, while the y-coordinate of target will be zero.

Running the program we find: $c = 3.698 m^2 kg$.

Example 8.3

Shooting with the 0.338 GB528 Lapua Scenar bullet in ICAO atmosphere.

(a) Find the departure angle α_{0T} needed to hit the target located at the zero range $x_T = 1800 m$. Initial velocity is 830m/s. BC = 3.796 corresponds to characteristic Lapua G-function.

(b) Find the super elevation angle α_{0i} needed to hit the target located at the inclined range $D_T = 1800m$ if the elevation angle is $E = 30°$.

(c) For the same bullet, use the reference G_7 - function to find the departure angle needed to hit the target at horizontal zero range $x_T = 1800m$.

Consider ICAO atmosphere and the ballistics coefficient given by Wikipedia: $C = 0.377 \ lb/in^2$. BC in metric unit is $c = 1.4222 / 0.377 = 3.772 m^2 / kg$.

Solution

Using Alapua16.BAS.
(a) $\alpha_{0T} = 1.601$, $v = 311.85 m/s$, $t = 3.747 s$. $c = 3.796$
(b) Input the inclined range $D_T = 1800m$ and the elevation angle $E = 30°$:

Executing the PC program we obtain the departure angle
$$\alpha_{0i} + E = 31.387°.$$

Since $E = 30°$, we find that super elevation angle is $\alpha_{0i} = 1.387°$.

Using the approximate formula (2.1.1), we find approximately the same value
$$\alpha_{OI} = \alpha_{0T} \cdot \cos(E) = 1.6069 \cdot \cos(30°) = 1.392°.$$

We can easily see that there is a difference of 0.28 minutes between two values of super elevation angle.

(c) We can use APROJ16.BAS.

Using APROJ16.BAS we find the departure angle $\alpha_{0T} = 1.649°$.

Comment

Comparing the approximate value $\alpha_{0T} = 1.649°$ with the exact value $\alpha_{0T} = 1.6069$ found in (a), we see that there is a difference of
$$\Delta\alpha_{0T} = 1.649 - 1.6069 = 0.0421° = 2.52'$$

Example 8.4

Shooting with M118LR bullet (Long Range, sniper bullet), mass $m = 0.01134 kg$, caliber $d = 0.0078232\ m$, departure velocity, $v_0 = 884\ m/s$, ballistic coefficient, $BC = 5.3970\ m^2/kg$ that corresponds to characteristic G-function B6, appendix B. Consider the ASM atmosphere.

(a) Find the departure angle α_{0T} needed to hit the target located at zero range $x_T = 1500m$.

(b). Find the super elevation angle α_{0i} needed to hit the target located at the inclined range $D_T = 1500m$ if the inclined plane is $E = 25°$.

(c) Solve case (b) when there is a tail-wind with velocity 5m/s.

Solution

Using PC program ACHA16.BAS.
Procedure:
Select ICAO = 2. Select M118LR = 5

Input:

X-coordinate of target = 1500. Y-coordinate of target = 0. Y-coordinate of firearm = 0. Departure velocity = 884. Temperature = 15. Temperature of propellant = 21.11. Pressure = 750. Humidity = 0.78. Projectile mass = 1 (any non-zero number if there is no change in mass). Change in mass = 0. Range-wind = 0. Cross wind = 0. BC = 5.3970.

Output:

Departure angle = 1.4992 degree. time = 3.332s. velocity 288.47m/s. Impact angle = -3.113 degree. Trajectory vertex has the coordinates (894m, 14.62m).
(a) Thus $\alpha_0 = 1.4992°$, $v = 288.49m/s$, $t = 3.331s$.

Note.

The departure angle predicted by Raymond Von Wahlde and Dennis Metz is $\alpha_0 = 1.4929°$, while the velocity and time are respectively $v = 288.31m/s$ and $t = 3.339s$.

(b) Input the inclined distance $D = 1500m$ and the elevation angle $E = 25°$ together with other elements given in the exercise.

Executing the PC program we obtain the departure angle

$$\alpha_{0i} + E = 26.3516°.$$

Since $E = 25°$ we find that super elevation angle is $\alpha_{0i} = 1.3516°$.

Using the approximate formula (2.1.1), we find the same value

$$\alpha_{OI} = \alpha_{0T} \cdot \cos(E) = 1.4929 \cdot \cos(25°) = 1.353°.$$

(c) Input the inclined distance $D = 1500m$, the elevation angle $E = 25°$, and the velocity of wind together with other elements given in the exercise.

Executing the PC program we obtain the departure angle

$$\alpha_{0i} + E = 26.2173°.$$

Hence we find that $\alpha_{0i} = 1.2173°$

Example 8.5

Shooting in ASM atmosphere with a **M118 Ball bullet** with velocity $v_0 = 792.48 \; m/s$. It is given the departure angle $\alpha_0 = 0.8993°$, (when elevation angle is zero)`.

Use the reference G_7-function and the corresponding BC, $c = 5.9258 \; m^2/kg$.

Consider that in ASM atmosphere, the temperature of air, the temperature of propellant of propellant, the pressure and humidity are respectively 15 degree Celsius, 21.11 degree C., pressure 750mm Hg, humidity 78% = 0.78.

Find:

(a) The coordinates of the point on the trajectory that correspond to the horizontal distance x =1000 meter.

(b) The coordinates of the point on the trajectory that correspond to horizontal range x = 1000m if there is a constant tail range-wind of +4.5m/s.

ELEMENTS OF EXTERIOR BALLISTICS

(c) Find the point of impact at the inclined range 1000m, if the marksman fires with the same super elevation angle $\alpha_0 = 0.8993°$ when the elevation angle of shooting site is +25 degree. Weather is without wind.

(d) Solve (c) if there is a range wind of +4.5m/s

Solution

Using RPROJ16.BAS and the data given above. Input 1000 as horizontal range. Use ASM option (Press 1) and G7.

We find:

(a) Bullet will hit at the point with coordinates $x_T = 999.9m$ and $y_T = 0.31m$.

(b) $x_T = 998.97m$, and $y_T = 0.044m$.

(c) $x_T = 906.23m$, $y_T = 424.77m$ Inclined range 1000.84m. The bullet will pass 2.19m over the center of the target located at incline range 1000m.

(d) $x_T = 906.29m$, $y_T = 427.40m$ Inclined range 1002m. The bullet will pass 4.788m over the center of target located at the inclined range 1000m.

Example 8.6 Projectile Drop

A marksman fires a **M118 Ball bullet**. The shooting site is 100 meters over the sea level. The initial velocity of bullet is $v_0 = 792.48\ m/s$, while BC of the given bullet with respect to the reference G_7 - function is $c = 5.9258\ m^2/kg$.

Consider that, the temperature of air and of propellant are equal to 10 degree Celsius, pressure 750mm Hg, humidity 50%.

Find the drop of projectile that correspond to the horizontal distance x=1200 meter.

(a) Select the ASM option (press 1).
(b) Select the ICAO option, (press 2).

Note that the y-coordinate of the muzzle is 100m.

Solution

We use RPROJ16.BAS and the data given above. Input 1200 as horizontal range. We find:

Running the PC program we find:

(a) The drop at 1200 meters is -26.78 meter.
(b) The drop at 1200 meters is -26.80 meter

The drop must be identical since the bullet is fired at the same shooting site and with the same data. Note that there is a small difference in the drop due to the numerical method of solution of differential equations.

Example 8.7 Declined Shooting

Consider the example 2.2, section 2.2.

Use the PC program Rlapua16.bas to find the elements of the Lapua GB528 19.44 bullet. Declined range $R_S = 1000$ yards, the depression angle $E = 30$ degree. Consider the ICAO atmosphere, the initial velocity 830m/s and the super elevation angle $\alpha_{OI} = 0.4523°$, estimated in example 2.2 section 2.2.

Solution

Input:

ICAO atmosphere, GB528 Lapua, declined (Press 2), declined distance $R_S = 1000 \cdot 0.9144 = 914.4m$, depression angle $E = 30°$, departure angle $\alpha_{OI} = 0.4523°$, y-coordinate of muzzle $y_0 = R_s \cdot \sin(E) = 914.4 \cdot \sin(30) = 457.20m$, departure velocity 830m/s, temperature of air = 15 degree Celsius, temperature of cartridge = 21.11 degree Celsius, pressure 760mm Hg., humidity = 0, projectile mass = 1 (or whatever number different from zero), change in projectile mass = 0, range wind = 0, cross wind = 0, BC = 3.796.

Step size in this case should be 0.1, or any other number but not zero. The real value of the bullet mass is required to be input if we are studying the influence of small changes in mass of the bullet.

Output: Coordinates of impact point (791, 85). The impact angle is $\alpha_T = -30.605°$.

The bullet will fall 0.89m before the center of the target located at the point with abscissa $x_T = R_s \cdot \cos(E) = 914.4 \cdot \cos(-30°) = 791.89m$, and $y_T = 0$.

The vertical deviation of the given bullet from the center of the target at the declined range 914.4m is $\Delta y = 0.272m$.

3

Standard Atmosphere in Exterior Ballistics

Introduction

The resistance of air to the flight of a projectile depends on the characteristics of atmosphere, on the physical and aerodynamic characteristics of projectile, and on the instantaneous velocity and Mach number.

Characteristics of Atmosphere

The characteristics of atmosphere are the density ρ, the temperature T, the pressure p, the humidity of air, and wind.

The exterior ballistics of point-mass projectile considers an ideal atmosphere where wind is not present while air density, the pressure and the air temperature decrease with increasing altitude over the sea level, according to the ideal gas law and the hydrostatic equation (section 3.3).

At the sea level, or close to it, for mid-altitude geographic locations, the characteristics of atmosphere are considered constant and unchanged with time and location. Such atmosphere is called Standard Atmosphere.

The standard atmosphere is a hypothetical atmosphere where density, temperature, pressure, humidity of air and speed of sound at the sea level (or close to it), as well as their vertical variation with altitude over the sea level are established by international agreements.

3.1 Standard Atmosphere

In the practice of exterior ballistics there are in use at least three standard atmospheres: the ICAO atmosphere, the ASM atmosphere, and TSA atmosphere.

ICAO Standard Atmosphere

The International Civil Aviation Organization atmosphere, (known as ICAO atmosphere) has the following characteristics:

- At the sea level, $y = 0$, the air density is $\rho_{0N} = 1.2251 \ kg/m^3$, the temperature of air is $t_{0N} = 15°Celsius$ ($T_{0N} = 288.15 \ Kelvin$), the atmospheric pressure is $p_{0N} = 760mmHg$, the relative humidity of air vapors is $r_H = 0\%$, the speed of sound is $a_{0N} = 340.30 m/s$.

- The corresponding virtual temperature is $\tau_{0N} = 288.15°K$.

- The wind is absent.

ASM Standard Atmosphere

The Army Standard Metro (known as ASM) atmosphere has the following characteristics:

- At the firing site ($y = 0$), the air density is $\rho_{0N} = 1.2034 kg/m^3$, the temperature of air is $t_{0N} = 15 \ °Celsius$, the atmospheric pressure is $p_{0N} = 750 \ mm \ Hg$, the relative humidity is $r_H = 78\%$, the speed of sound is $a_{0N} = 341.458 \ m/s$.

- The corresponding virtual temperature is approximately $\tau_{0N} = 289.60°K$.[15]

- The wind is absent.

TSA Standard Atmosphere

The Traditional Standard Atmosphere (TSA) mostly used in ex-eastern countries, assumes that the projectile motion is in a standard atmosphere, with the following characteristics:

- At the firing site ($y = 0$) the air density is $\rho_{0N} = 1.205 \ kg/m^3$, the temperature of air is $t_{0N} = 15°Celsius$, the virtual temperature is $\tau_{0N} = 289.08 K$, and corresponds to the relative humidity $r_H = 50\%$, the atmospheric pressure is $p_{0N} = 750 \ mm \ Hg.$, the speed of sound is $a_{0N} = 340.84 \ m/s$.

- The wind is absent.

The TSA atmosphere assumes that the projectile is fired at an altitude of 110 meters, over the sea level, where the atmospheric pressure is considered 750 mm Hg.

Note

Bullet companies in USA use ICAO, or ASM atmosphere.
Thus, Sierra, Hornady and Barnes use the ASM atmosphere.
Lapua and Berger use ICAO atmosphere.
The Russian and other ex-East European countries use the TSA atmosphere.

3.2 Characteristics of Atmospheric Air

In a standard atmosphere, the air is considered an ideal gas at rest that follows the ideal gas law and the hydrostatic equation that describe the decrease of pressure with altitude, i.e. the equations

$$\rho = (1 - 0.3785 \frac{e}{p}) \frac{\mu_A}{R} \frac{P}{T}, \tag{3.2.1}$$

$$p = (R/\mu_A) \cdot \rho \cdot \tau, \tag{3.2.2}$$

$$dp/dy = -\rho \cdot g, \tag{3.2.3}$$

where $R = 8.31441 J \cdot mol^{-1} K^{-1}$, $\mu_A = 28.9644 \cdot 10^{-3} kg/mol^{-1}$ are respectively the universal gas constant and the molar mass of air, while T is the temperature of air in degree Kelvin, τ is the virtual temperature of air (degree Kelvin), ρ is the density of air (kg/m³), and p is the atmospheric pressure in Pa.

$(1 Pa = 1 N/m^2 = 7.5006 \cdot 10^{-3} mm\ Hg$, $1\ mm\ Hg = 133.322\ Pa.)$

Virtual Temperature, Relative Humidity and Vapor Pressure

The virtual temperature τ can be estimated using equation

$$\tau = \frac{T}{1 - 0.3785 \cdot (e/p)}, \qquad (3.2.4)$$

where T is the temperature of dry air in degree Kelvin,

$$T = 273.15 + t, \qquad (3.2.5)$$

e is the partial pressure of saturated water vapor, when relative humidity has a certain percentage value, r_H.

The parameter t is the temperature of dry air in degree Celsius.

The partial pressure e and the atmospheric pressure p must be expressed in the same units, i.e. both in Pascal (Pa), both in mm Hg, or inch Hg.

Employing (3.2.4), the density of humid air (3.2.1) can be expressed through the virtual temperature by the equation

$$\rho = \frac{\mu_A}{R} \frac{P}{\tau}. \qquad (3.2.6)$$

Partial Vapor Pressure

The saturated pressure of water vapor that corresponds to relative humidity $r_H = 100\%$, can be calculated using the following approximate formulas[16]:

$$e_{100\%} = 7.50187 \cdot e^{19.04 \cdot (1. - 280.07/T)}, \quad 273.16 \leq T \leq 327.15, \qquad (3.2.7)$$

or

$$e_{100\%} = 7.50187 \cdot e^{22.024 \cdot (1.-279.24/T)}, \quad 255.15 \leq T \leq 273.15, \quad (3.2.8)$$

where T is the temperature of dry air ($r_H = 0.00\%$) in degree Kelvin.

Multiplying (3.2.7), or (3.2.8) by the relative humidity, r_H, we obtain the corresponding vapor pressure for a given temperature and a given relative humidity.

In standard atmosphere at the sea level, or at shooting ground The virtual temperature τ_{0N}, can be estimated using (3.2.4):

$$\tau_{0N} = \frac{T_{0N}}{1 - 0.3785(e_{0N} / p_{0N})}, \quad (3.2.9)$$

where e_{0N} is the partial pressure of water vapor at the standard temperature $t_{0N} = T_{0N} - 273.15$ and relative humidity r_H.

The saturated pressure of water vapor e can be estimated as well using the data presented in the following table 2.1 (temperature t in degree Celsius, pressure of vapor e in mm. Hg).

Table 2.1 – Saturated Pressure of Water Vapor and (relative humidity, $r_H = 100\%$)

Temp. °C	-40	-18	-10	0.0	5.0	10	15
Pressure e	0.15	1.14	1.95	4.58	6.54	9.20	12.70
Temp. °C	20	25	30	38	40	54	-
Pressure e	17.54	23.76	31.7	49.2	55.1	115.1	-

If the relative humidity r_H is less than 100%, for example it is $r_H = 50\%$, then, the partial pressure e of vapor that corresponds to relative humidity $r_H = 50\%$ is obtained by multiplying the respective value in table 2.1 with $r_H = 50\% = 0.50$. Thus, at the temperature of air $t_0 = 15°Celsius$ (288.15

Kelvin), and $r_H = 50\%$ humidity, the partial pressure of water vapor is $e = 6.35 mmHg$.

Substituting the above value in (3.2.4), we find that the virtual temperature is 289.08 Kelvin.

The partial pressure e of the saturated water vapor that corresponds to an intermediate temperature that is not displayed in table 2.1 can be found by interpolation.

Speed of Sound

Speed of sound depends as well on the temperature and humidity of air.

The speed of sound a, for a given virtual temperature τ, can be estimated using the equation[17]:

$$a = 20.0469\sqrt{\tau}. \qquad (3.2.10)$$

Example 2.1

The relative humidity of air is $r_H = 78\% = 0.78$, while the temperature and atmospheric pressure are respectively $t_0 = 15°Celsius$ and $p_0 = 750\ mm\ Hg$.

Find the virtual temperature, density of air, and the speed of sound.

(Note that $1\ Pa = 7.5006 \cdot 10^{-3} mm\ Hg$; $1\ mm\ Hg = 133.322\ Pa$.)

Solution

Employing equation (3.2.7), we find that the partial pressure of water vapor is approximately

$$e = r_H \cdot e_{100\%} = (0.78) \cdot 7.50187 \cdot e^{19.04 \cdot (1 - 280.07/(273.15+15))}$$
$$= 9.980\ mm\ Hg$$

The virtual temperature is
$$\tau = \frac{T}{1-0.3785(e/p)} = \frac{273.15+15}{1-0.3785(9.980/750)} = 289.61°Kelvin$$

Using the data of table 2.1, we find the same partial pressure and virtual temperature.

Indeed, table 2.1 shows that at temperature $t_0 = 15°Celsius$ the pressure of saturated water vapor is 12.7 mm Hg.

Thus, the pressure of vapor that corresponds to relative humidity 78% is
$$e = r_H \cdot e_{100\%} = (0.78) \cdot (12.7) = 9.906 \; mm \; Hg.$$

The virtual temperature is
$$\tau = \frac{T}{1-0.3785(e/p)} = \frac{273.15+15}{1-0.3785(9.906/750)} = 289.61°Kelvin$$

The density of air is
$$\rho = \frac{\mu_A}{R} \frac{P}{\tau} = \frac{28.9644 \cdot 10^{-3}}{8.31441} \cdot \frac{750 \cdot (133.322)}{289.61} = 1.203 \; kg/m^3.$$

The speed of sound is
$$a = 20.0469\sqrt{\tau} = 20.0469 \cdot \sqrt{289.61} = 341.16 \; m/s.$$

Example 2.2 Find the virtual temperature, the density of air, and the speed of sound at a firing site where the pressure, the temperature of dry air, and the relative humidity are respectively 760 mm Hg, 25 degree Celsius, and 65%.

Solution

Employing Lewis' Formula (3.2.7) we find that the pressure of water vapors is
$$e = (0.65) \cdot (7.50187 \cdot e^{19.04 \cdot (1-280.07/(273.15+25))}) = 15.471 \; mm \; Hg.$$

The virtual temperature is
$$\tau = \frac{T}{1-0.3785(e/p)} = \frac{273.15+25}{1-0.3785(15.471/760)} = 300.47 \; °Kelvin$$

The density of humid air is

$$\rho = \frac{\mu_A P}{R \tau} = \frac{28.9644 \cdot 10^{-3}}{8.31441} \cdot \frac{(760) \cdot 133.322}{300.47} = 1.1748 \ kg/m^3.$$

The speed of sound is

$$a = 20.0469\sqrt{\tau} = 20.0469 \cdot \sqrt{300.47} = 347.491 \ m/s.$$

Example 2.3

(a) Find the speed of sound that corresponds to the ICAO atmosphere at the sea level.

(b) Find as well the speed of sound at the ground for the ASM atmosphere.

Consider the shooting ground at the altitude 110 meters above the sea level.

Solution

For the ICAO atmosphere, at the sea level, relative humidity is zero, $r_H = 0$. Using (3.2.8), it results that the corresponding pressure of vapor is zero, $e = 0$. Substituting $e = 0$ in (3.2.9), we find that the virtual temperature is

$$\tau_{0N} = T_{0N} = 288.15 \ °K.$$

The speed of sound is

$$a = 20.0469\sqrt{\tau} = 20.0469\sqrt{288.150} = 340.30 m/s.$$

For the ASM standard atmosphere, at the firing ground 110 meters over the sea level, we have: $t_{0N} = 15 \ °Celsius$, $p_{0N} = 750$, $r_H = 78\%$.

At the temperature $t_{0N} = 15 \ °$, the pressure of saturated water vapor is 12.7 mm Hg. Thus, for the partial pressure of water vapor we find:

$$e = (0.78) \cdot (12.7) = 9.906 \ mmHg.$$

The virtual temperature is

$$\tau = \frac{T}{1 - 0.3785(e/p)} = \frac{288.15}{1 - 0.3785 \cdot (9.906/750)} = 289.60 \ °Kelvin.$$

The speed of sound is

$$a = 20.0469\sqrt{\tau} = 20.0469\sqrt{289.60} = 341.15 m/s.$$

Example 2.4 Fictive BC

The 0.308 M118LR bullet is launched with initial velocity 884m/s in a hot summer day, at a shooting site where the temperature of air is 35 degree Celsius, air pressure is 755 mm Hg., humidity is 75%.

Find the fictive ballistic coefficient of the given bullet $\overline{c}$, considering that the standard atmosphere is the ICAO atmosphere.

The BC of the given bullet with respect to the corresponding personalized G-function is $c_0 = 5.3970 m^2/kg$.

Solution

The characteristics of ICAO atmosphere at the sea level are:

Pressure, $p_0 = 760 mm\ Hg.$, relative humidity, 0%, speed of sound, $a_0 = 340.30\ m/s$, temperature, $t_0 = 15°C$; air density $\rho_0 = 1.225\ kg/m^3$.

The virtual temperature in ICAO atmosphere is

$$\tau_0 = T_0 = 273.15 + 15 = 288.15°. \quad (1)$$

Virtual temperature at the shooting site

Using table 2.1, section 3.2, by interpolation we find that the vapor pressure at 35 degree and humidity 100% is 42.64 mm Hg.

The vapor pressure, when humidity is 75%, is

$$e = 0.75 \cdot 42.64 = 31.98\ mm\ Hg.$$

The virtual temperature at firing site is

$$\tau_0 = \frac{T_0}{(1 - 0.3785 \cdot e_0/p_0)} = \frac{273.15 + 35}{(1 - 0.3785 \cdot (31.978)/755)} = 313.171°$$

Using equation (4.1.5), we find that the fictive BC, at the given site, is

$$\overline{c} = J^{-1} \cdot c = (\frac{p_0}{p_{0N}}\sqrt{\frac{\tau_{0N}}{\tau_0}}) \cdot c = \frac{755}{760} \cdot \sqrt{\frac{288.15}{313.171}} \cdot 5.3970 = 5.143.$$

The value of the scaling factor J^{-1} is

$$J^{-1} = (\frac{p_0}{p_{0N}}\sqrt{\frac{\tau_{0N}}{\tau_0}}) = \frac{755}{760} \cdot \sqrt{\frac{288.15}{313.171}} = 0.9529$$

while its inverse is

$$J = 1/J^{-1} = 1/0.9529 = 1.0494.$$

3.3 Change of Air Characteristics with Altitude

The decrease in the characteristics of standard atmospheric air, i.e. the decrease in the virtual temperature τ, density ρ, and pressure p with altitude y over the sea level (until around 11,000 meters) is described respectively by the following equations:

$$\tau = \tau_0 - 0.006328 y, \quad (3.3.1)$$

$$\rho = \rho_0 (\frac{\tau_0 - 0.006328 y}{\tau_0})^{4.4}, \quad (3.3.2)$$

and

$$p = p_0 (\frac{\tau_0 - 0.006328 y}{\tau_0})^{5.4}, \quad (3.3.3)$$

where τ_0, ρ_0, and p_0 are respectively the virtual temperature, the density, and the pressure of air at the firing site, above or at the sea level.

The number -0.006328 is the vertical temperature gradient,
$d\tau / dy = -0.006328$.

As a matter of fact, International Standard Atmosphere (ISA), and ICAO atmosphere considers a temperature gradient $d\tau / dy = -0.0065$.

Relative Density of Air

The relative density of air is the ratio of the density of air at a given altitude over the standard density of air at the sea level, i.e.

$$h(y) = \rho / \rho_{0N}. \qquad (3.3.4)$$

Using the equation of (3.3.2), we find that the relative density of air, at the altitude y is

$$h(y) = (\frac{\tau_{0N} - 0.006328 \cdot y}{\tau_{0N}})^{4.4}. \qquad (3.3.5)$$

Empirical Formulae

For thee relative density of air can be used the following empirical formulas:

$$h(y) = e^{-0.0001 \cdot y}, \qquad (3.3.6)$$

$$h(y) = 1 - 0.0001 \cdot y, \qquad (3.3.7)$$

Example 3.1

The temperature of dry air, the pressure and the humidity, at the firing site, are respectively 20 degree Celsius, 750 mm Hg and 75%.

Find the virtual temperature, the density, and the pressure of air at the altitude 600 meters over the firing site.

Solution

At the firing site, we find:
The pressure of water vapors is

$$e = (0.75) \cdot (7.50187 \cdot e^{19.04 \cdot (1 - 280.07/(273.15+20))}) = 13.158 \ mm \ Hg.$$

The virtual temperature is

$$\tau = \frac{T}{1 - 0.3785(e/p)} = \frac{273.15 + 20}{1 - 0.3785(13.158/750)} = 295.11 \ °Kelvin.$$

ELEMENTS OF EXTERIOR BALLISTICS | 101

The virtual temperature and the relative density of air, at the altitude 600 meters, over the ground, are respectively:

$$\tau = \tau_0 - 0.006328 y = 295.11 - 0.006328 \cdot (600) = 291.31 \,°K.$$

and

$$h(y) = (\frac{\tau_{0N} - 0.006328 y}{\tau_{0N}})^{4.4} = (\frac{295.11 - 0.006328 y}{295.11})^{4.4} = 0.9446.$$

The pressure at altitude 600 m, is

$$p = p_0 (\frac{\tau_0 - 0.006328 y}{\tau_0})^{5.4} = 750 \cdot (\frac{295.11 - 0.006328 \cdot (600)}{295.11}) = 699.347 \; mm \; Hg.$$

Example 3.2

At sea level, the pressure is 760 mm Hg, while the temperature is 15 degree Celsius (288.15 degree Kelvin).

At what altitude we expect to have a pressure of 750 mm Hg?

Solution

Substituting in (3.3.3), i.e. in

$$p = p_0 (\frac{\tau_0 - 0.006328 y}{\tau_0})^{5.4},$$

we can write:

$$760 \cdot (\frac{288.15 - 0.006328 y}{288.15})^{5.4} = 750.$$

Hence, solving for y, we find $y = 111.55 \; m.$

Example 3.3

The temperature and pressure, measured at a meteorological station located 500 meters above the sea level, are respectively $10°\,C$ and 710 mm Hg, while the relative humidity is 70%.

Find the atmospheric pressure, the virtual temperature, the temperature, and the relative humidity of air at the sea level.

Solution

The temperature measured at the station location in Kelvin is
$$T = 273.15 + 10° = 283.15 K.$$

Employing equation (3.3.7) we find that the pressure of water vapors is
$$e_{70\%} = (0.70) \cdot 7.50187 \cdot e^{19.04 \cdot (1.-280.07/T)} =$$

$$= 7.50187 \cdot e^{19.04 \cdot (1-280.07/283.15)} = 6.46 \; mm \; Hg.$$

The corresponding virtual temperature is
$$\tau = \frac{T}{1 - 0.3785 \cdot (e/p)} = \frac{283.15}{1 - 0.3785 \cdot (6.46/710)} = 284.13°K.$$

The density of humid air is
$$\rho = \frac{\mu_A}{R} \frac{P}{\tau} = \frac{28.9644 \cdot 10^{-3}}{8.31441} \cdot \frac{(710) \cdot 133.322}{284.13} = 1.161 \; kg/m^3.$$

The temperature, the density, and the pressure of air at the sea level:

Employing (3.3.2), we find that the virtual temperature at the sea level is
$$\tau_0 = \tau + 0.006328 \cdot y = (284.13 + 0.006328 \cdot (500))$$
$$= 287.29.°K = 14.14 \; °C.$$

Using (3.3.2) and (3.3.3), for the density of atmospheric air, and the pressure at the sea level, we find respectively:
$$\rho_0 = \rho(\frac{\tau_0}{\tau})^{4.4} = 1.161 \cdot (\frac{287.29}{284.13})^{4.4} = 1.173 \; kg/m^3,$$

and
$$p_0 = p(\frac{\tau_0}{\tau})^{5.4} = 710 \cdot (\frac{287.29}{284.13})^{5.4} = 753.69 \; mm \; Hg.$$

3.4 Barometric Formula

The Isothermal Barometric Formula is an approximate equation that expresses the change in pressure with altitude above the sea level, or above a shooting ground, not necessary located at the sea level.

It is obtained assuming that the gravity is constant and the atmosphere is isothermal, i.e. that the atmospheric temperature of air T does not change with altitude (temperature gradient is zero, $dT/dy = 0$).

The Isothermal Barometric Formula can be used when the maximum altitude of the projectile trajectory is relatively small, so that the temperature of air can be considered constant for that layer of air the projectile moves into.

Thus, the barometric formula can be used in long range shootings where the trajectory height is relatively small.

The change of pressure with altitude can be approximately estimated using the isothermal barometric formula[18]:

$$p = p_0 \cdot e^{-\mu_A \cdot g \cdot y/(RT)}, \qquad (3.4.1)$$

where p_0 is the pressure at the firing ground, p is the pressure at an altitude "y" over the ground, μ_A is the molar mass of air ($\mu_A = 28.9644 \cdot 10^{-3}$ kg/mol⁻¹), R is the universal constant of ideal gas $R = 8.31441 \ J \cdot mol^{-1} K^{-1}$, g is gravitational acceleration, $g = 9.80665 m/s^2$ and T (assumed constant) is the temperature of dry air in Kelvin.

Substituting in (3.4.1), $\mu_A = 28.9644 \cdot 10^{-3}$, $R = 8.31441 \ J \cdot mol^{-1} K^{-1}$, $g = 9.80665 m/s^2$, we can write:

$$p = p_0 \cdot e^{-0.03416 \cdot y/T}. \qquad (3.4.2)$$

The data for the weather, including the atmospheric pressure, are provided by a meteorological station that in general is not located at the firing ground.

The measured data, including the pressure at the altitude of the station location, are reported to the shooting ground, or the measured pressure is brought to the sea level.

Based on the measured pressure p_s at the station location (altitude y_s), we find the pressure p_b at the altitude y_b where the firearm is located.

Using the barometric formula (3.4.2), we can write respectively

$$p_b = p_0 \cdot e^{-0.03416 \cdot y_b / T},$$

and

$$p_s = p_0 \cdot e^{-0.03416 \cdot y_s / T}.$$

Dividing the above formulas, for the pressure at the firing site we can write:

$$p_b = p_s \cdot e^{-0.03416 (y_b - y_s)/T}. \tag{3.4.3}$$

A similar formula expresses the change of air density with altitude.

Indeed, since at a constant temperature the density is proportional to pressure, considering formulas (3.4.2) and (3.4.3), we can write:

$$\rho = \rho_0 \cdot e^{-0.03416 \cdot y / T}.$$

and

$$\rho_b = \rho_s \cdot e^{-0.03416 (y_b - y_s)/T}$$

Since the coefficient $-0.03416/T$ is relatively small we can approximate

$$p_b \approx p_s (1 - 0.03416 \cdot \frac{y_b - y_s}{T}) \tag{3.4.4}$$

Once the pressure is calculated using (3.4.14), we can use the ideal gas equation

$$\rho = \frac{\mu \cdot p}{R \cdot T} = 0.003484 \frac{p}{T}. \tag{3.4.5}$$

to find the corresponding density of dry air.

The quantity $h = T / 0.03416$ in the right side of above formula is called scale altitude. Scale altitude when we consider the standard temperature 15 degree Celsius (288.15 Kelvin) is $h = T / 0.03416 = 288.15 / 0.03416 = 8435.30m$.

Practical rule to estimate the atmospheric pressure

In practice, to estimate the pressure of air at a given altitude, when it is known the atmospheric pressure at the shooting ground, or at the altitude of meteorological station, we can use the following practical rule.

Considering that the atmospheric pressure decreases approximately 8.50 mm Hg for any 100 meters increase in the altitude, the pressure at an altitude y can be estimated approximately by the formula:

$$p_y = p_0 - 8.5 \cdot (y - y_0)/100, \qquad (3.4.6)$$

where p_0 is the pressure at the altitude y_0.

Thus, if the pressure p_0 at the sea level is 760 mm Hg then at the altitude 100 meters over the sea level it is 751 mm Hg, while at the altitude 1000 meters the pressure is approximately

$$p_y = p_0 - 8.5 \cdot (y - y_0)/100 = 750 - 8.5 \cdot (1000 - 0)/100$$
$$= 665 \; mm \; Hg.$$

Note

In practice long range shooting, the best way to estimate pressure, temperature is to measure those parameters at the firing site.

Example 4.1

The temperature and pressure, measured at a meteorological station located 500 meters above the sea level, are respectively $10°\,C$ and 710 mm Hg, while the relative humidity is 70%.

Find the atmospheric pressure, the virtual temperature, the temperature, and the density of air at the sea level.

Solution

The temperature measured at the station location in Kelvin is

$$T = 273.15 + 10° = 283.15 K.$$

Since the atmosphere is isothermal, the same temperature is assumed at the sea level.

At sea level, the pressure is
$$p_b = p_s \cdot e^{-0.03416(y_b - y_s)/T} = 710 \cdot e^{[-0.03416 \cdot (0-500)/(283.15)}$$
$$= 754.15 \ mm \ Hg.$$

The pressure is quite equal to the value we calculated in example 3.6 (above).

Using (3.4.4) we get approximately the same result:
$$p_b \approx p_s (1 - 0.03416 \cdot \frac{y_b - y_s}{T}) = 710 \cdot (1 - 0.03416 \cdot \frac{0-500}{273.15+10}) = 752.83 \ mm \ Hg$$

The virtual temperature at sea level:

The temperature of air at sea level is considered equal to the station temperature since the atmosphere is considered isothermal.

Employing equation (3.3.7) we find that the pressure of water vapors at sea level is
$$e_{70\%} = (0.70) \cdot 7.50187 \cdot e^{19.04 \cdot (1. - 280.07/T)} =$$

$$= 7.50187 \cdot e^{19.04 \cdot (1 - 280.07/283.15)} = 6.46 \ mm \ Hg.$$

The virtual temperature is
$$\tau = \frac{T}{1 - 0.3785 \cdot (e/p)} = \frac{283.15}{1 - 0.3785 \cdot (6.46/754.15)} = 284.07°K$$

The density at sea level the density is
$$\rho = \frac{\mu_A}{R} \frac{p}{\tau} = \frac{28.9644 \cdot 10^{-3}}{8.31441} \cdot \frac{(754.15) \cdot 133.322}{284.07} = 1.233 \ kg/m^3$$

Note

We find approximately the same value of air pressure using the approximate formula (3.4.6):
$$p = 710 - 8.5 \cdot (0 - 500)/100 = 752.50 \ mm \ Hg.$$

4

Elementary Exterior Ballistics

Introduction

Elementary equations describe the projectile trajectory in standard and non standard atmosphere. They are approximate equations but the prediction accuracy of the elements of the ballistics trajectory we obtain employing those equations is remarkable.

Following Shapiro, we estimate the Coriolis Effect.

The cross-wind drift is predicted using Didion's elementary equation.

We also describe methods that allow the long range shooter to measure in practice of shooting, the wind drift, the spin drift, and the Coriolis Effect.

We show the remarkable Newton-Snell's law and tangent law on similar trajectories that are some elementary techniques to predict accurately the elements of bullet trajectory in a non-standard atmosphere when we know the standard range table.

Particularly, those two laws give the possibility to the shooter to easily calculate the departure angle, as well as the aiming angle to zero in the rifle in non standard atmosphere.

The elementary equations can be used for the horizontal shooting or for the inclined shooting.

4.1 Modified Piton-Bressant Trajectory

The simplest parabola that describes the trajectory of projectile flight in presence of resistance of air, but the least accurate, is the 3rd degree parabola that is known also as Piton-Bressant trajectory equation:

$$y = \tan \alpha_0 \cdot x - \frac{gx^2}{2v_0^2 \cos^2 \alpha_0} \cdot (1 + \frac{A_1 \cdot x}{\cos \alpha_0}), \quad (4.1.1)$$

Piton-Bressant parabola is shown in Exterior Ballistics with Applications, 3rd. edition, 2011.

To improve the prediction accuracy of the trajectory, in the equation of Piton-Bressant (4.1.1) we introduce a factor "b" that can be determined experimentally together with the constant factor A_1.

The Piton-Bressant third order parabola trajectory in non-standard atmosphere is[19]

$$y = \tan \alpha_0 \cdot x - \frac{gx^2}{2v_0^2 \cos^2 \alpha_0} b \cdot (1 + \frac{J^{-1} A_1 \cdot x}{\cos \alpha_0}), \quad (4.1.2)$$

where b and A_1 are parameters that can be estimated using field shooting tests or reliable standard range tables, possibly Doppler radar data.

We have included also the scaling factor J^{-1},

$$J^{-1} = \frac{p_0}{p_{0N}} \sqrt{\frac{\tau_{0N}}{\tau_0}},$$

when shooting site is not in a standard atmosphere.

The other equations associated with modified Piton-Bressant equation (4.1.2) are:

Angle of flight:

Angle of flight α at any point on the trajectory with abscissa x, or at the impact point x_T is

$$\tan \alpha = \tan \alpha_0 - \frac{gx}{v_0^2 \cos^2 \alpha_0} b \cdot (1 + \frac{3}{2} \frac{J^{-1} A_1}{\cos \alpha_0} x), \quad (4.1.3)$$

Projectile Velocity

$$v = \frac{v_0 \cos \alpha_0}{\cos \alpha} b^{-1/2} \cdot (1 + 3 \frac{J^{-1} A_1 \cdot x}{\cos \alpha_0})^{-1/2}. \tag{4.1.4}$$

Time of Flight

$$t = \frac{b^{1/2}}{v_0 \cos \alpha_0} \int_0^x (1 + 3J^{-1} A_1 x / \cos \alpha_0)^{1/2} dx. \tag{4.1.5}$$

Departure angle

$$\tan \alpha_0 = \frac{gx}{2v_0^2 \cos^2 \alpha_0} b \cdot (1 + \frac{J^{-1} A_1 \cdot x}{\cos \alpha_0}), \tag{4.1.6}$$

or

$$\sin(2\alpha_0) = \frac{gx}{v_0^2} b \cdot (1 + \frac{J^{-1} A_1 \cdot x}{\cos \alpha_0}), \tag{4.1.7}$$

Note that if the projectile is launched in standard atmosphere then $J^{-1} = 1$.

The trajectory equation (4.1.2) and the related equations are easy to be used in practice of long range shooting.

For the inclined shooting, the departure angle is obtained by substituting in (4.1.2) the coordinates x and y of the impact point, and then solving the resulting equation for the departure angle (See for analogy example 4.2).

Projectile drop

From trajectory equation (4.1.2), we find the projectile drop, $\bar{y} = y - \tan \alpha_0 \cdot x$

$$\bar{y} = -\frac{gx^2}{2v_0^2 \cos^2 \alpha_0} b \cdot (1 + \frac{J^{-1} A_1 \cdot x}{\cos \alpha_0}), \tag{4.1.8}$$

At the impact point, the horizontal range is $x = x_T$ and the y-coordinate is $y_T = 0$.

Shooting with Departure Angle Equal to Zero

The Piton-Bressant equation and all related formulae can be simplified if the departure angle is zero. Substituting $\alpha_0 = 0°$ in the above equations we have:

Equation of parabola

$$y = -\frac{gx^2}{2v_0^2} \cdot b \cdot (1 + J^{-1} A_1 x), \quad (4.1.9)$$

Note that y is the projectile drop, i.e. $y = \bar{y}$.

Angle of Flight

$$\tan \alpha = -\frac{gx}{v_0^2} b \cdot (1 + \frac{3}{2} J^{-1} A_1 x), \quad (4.1.10)$$

Projectile Velocity

$$v = \frac{v_0}{\cos \alpha} b^{-1/2} \cdot (1 + 3 \cdot J^{-1} A_1 x)^{-1/2}. \quad (4.1.11)$$

Time of Flight

$$t = \frac{b^{1/2}}{v_0 \cos \alpha_0} \int_0^x (1 + 3 J^{-1} A_1 x / \cos \alpha_0)^{1/2} dx. \quad (4.1.12)$$

Using (4.1.2), or (4.1.9) we can find the parameters b and A_1.

Example 1.1

For the Lapua GB528 Scenar 19.4 g, Caliber 8.6 mm bullet find b and A_1 considering that shooting is done in Standard ICAO Atmosphere ($J^{-1} = 1$).

Departure angle: $\alpha_0 = 0$. Use the following data:
Range $x_1 = 900m$, drop $\bar{y}_1 = -8.14m$. Range $x_2 = 1200m$, drop $\bar{y}_2 = -16.571m$.

Solution

Substituting in (4.1.9) we have:

$$-8.14 = -\frac{g(900)^2}{2(830)^2} \cdot b \cdot (1+(1)A_1(900)), \tag{1}$$

and

$$-16.571 = -\frac{g(1200)^2}{2(830)^2} \cdot b \cdot (1+(1)A_1(1200)), \tag{2}$$

Dividing we find that

$$A_1 = 8.56596368 \times 10^{-4} \tag{3}$$

Substituting (3) in (2), we find

$$b = 0.79726343 \tag{4}$$

Thus the trajectory equation (4.1.2), for Lapua bullet, can be written:

$$y = \tan\alpha_0 \cdot x - \frac{gx^2}{2v_0^2 \cos^2\alpha_0}(0.79726)\cdot(1 + \frac{J^{-1}(8.56596\times 10^{-4}\cdot x)}{\cos\alpha_0}), \tag{5}$$

while (4.1.9) is

$$y = -\frac{gx^2}{2v_0^2}\cdot(0.79726)\cdot(1 + J^{-1}(8.56596\times 10^{-4})\cdot x), \tag{6}$$

In the same way we can write all the other equation.

Example 1.2

For the Lapua GB528 Scenar 19.44 g, Caliber 8.59mm, estimate:

(a) The drop of the bullet at horizontal range $x_T = 1000$ meters

(b) The angle that zeroes the firearm at 1000 meters.

Consider ICAO atmosphere and the initial velocity of bullet $v_0 = 830 m/s$.

Solution

(a) Using (4.1.9) and the parameters A_1 and b, found in example 1.1, we find the projectile drop

$$\bar{y} = -\frac{gx^2}{2v_0^2} \cdot b \cdot (1 + J^{-1}A_1 x) =$$

$$= -\frac{9.80665 \cdot (1000)^2}{2 \cdot (830)^2} \cdot (0.79726) \cdot [1 + (8.56596 \times 10^{-4}) \cdot (1000)] = 10.908m \, ,$$

Note that the drop estimated using differential equations, presented in chapter five is $\bar{y} = -10.439m$.

(b) **First method**
Using the non rigidity principle, we find that the departure angle needed to zero the firearm at 1000 meter is

$$\alpha_0 = \frac{\bar{y}}{x_T} \cdot \frac{180}{\pi} = \frac{10.908}{1000} \cdot \frac{180}{\pi} = 0.625°.$$

Second method
We can find the departure angle using (4.1.6) or (4.1.7).
Substituting we have:

$$\sin(2\alpha_0) = \frac{gx}{v_0^2} b \cdot (1 + \frac{J^{-1}A_1 \cdot x}{\cos \alpha_0}) = \frac{9.80556}{830^2} \cdot 0.79726 \cdot (1 + \frac{8.56596 \times 10^{-4} \cdot 1000}{\cos \alpha_0}) \quad (1)$$

We can easily solve the above equation using a graphing calculator.

Another way to solve the equation is to consider the value of departure angle, on the right side equal to zero, $\alpha_0 = 0$. Thus,

$$\sin(2\alpha_0) = \frac{9.80665 \cdot (1000)}{830^2} \cdot (0.79726) \cdot (1 + \frac{8.56596 \times 10^{-4} \cdot 1000}{\cos(0)}) = 1.20709 \, .$$

Solving the above equation we obtain an approximate value:

$$\alpha_0 = \arctan(1.20709)/2 = 0.60354°. \qquad (2)$$

Substituting $\alpha_0 = 0.60354$ on the right side of (1) and solving the resulting equation, we get the same value, $\alpha_0 = 0.60354°$.

Note that the value of the departure angle calculated in (2) is close to the value $\alpha_0 = 0.5958°$ obtained solving differential equations (5.1.2).

Substituting in (4.1.8), we find the bullet drop:

$$\bar{y} = -\frac{gx^2}{2v_0^2 \cos^2 \alpha_0} \cdot b \cdot (1 + \frac{J^{-1}A_1 \cdot x}{\cos \alpha_0}) =$$

$$= -\frac{9.80665 \cdot (1000)^2}{2 \cdot (830)^2 \cos(0.60354)} \cdot (0.79726) \cdot [1 + \frac{8.56596 \times 10^{-4}(1000)}{\cos(0.60354)}]$$

$$= -10.536m$$

Note that the estimated drop $\bar{y} = -10.536m$ is quite equal to $\bar{y} = -10.439m$ obtained using numerical integration of differential equations.

The following table 1.1 shows the elements of the Lapua GB528 Scenar bullet trajectory predicted using the equations (4.1.2) - (4.1.8) and the parameters (3) and (4), determined in example 1.1.

Table 1.1 Predicted Data Using Modified Piton-Bressant equations

Range (m)	0	300	600	900	1,200
Velocity (m/s)	830	698.26	583.34	511.22	460.58
Time (s)	0.000	1.404	0.852	1.3937	2.0435
Drop (m)	0.000	-0.644	-3.098	-8.146	-16.571

Note In all exercises it was assumed that the temperature of the black powder (projectile cartridge) is 21.11 degree Celsius. That means that the initial velocity is equal to standard velocity of Lapua projectile $v_0 = 830m/s$.

4.2 Exponential Equation of Projectile Trajectory[20]

The trajectory of point-mass projectile can be described relatively accurately employing exponential equation

$$y = \tan\alpha_0 \cdot x - \frac{g}{2v_0^2}\left(\frac{x}{\cos\alpha_0}\right)^2 e^{kx/\cos\alpha_0}, \qquad (4.2.1)$$

where "k" is a parameter that depends on the projectile and the coordinates of the point of impact, or the horizontal range.

The parameter k can be estimated using standard range tables, or shooting tests.

In a given standard atmosphere (ICAO, ASM, or TSA), the parameter "k" can be determined using the experimentally obtained coordinates of the point of impact (x_T, y_T) that correspond to a given departure angle α_0 ($\alpha_0 \neq 0$).

Substituting x_T and y_T, respectively for x and y in (4.2.1), we find the parameter "k":

$$k = \frac{\ln[y_T - x_T \cdot \tan\alpha_0) \cdot 2 \cdot v_0^2 \cdot \cos^2\alpha_0) \cdot (9.80665)^{-1} \cdot x_T^{-2}]}{x_T / \cos(\alpha_0)}. \qquad (4.2.2)$$

Note that the projectile drop at range x_T, (point on trajectory with coordinates $(x_T,\ )$ is

$$\bar{y} = y_T - x_T \tan\alpha_0. \qquad (4.2.3)$$

The other elements of the exponential trajectory are:

Departure angle that corresponds to zero range $(x_T, y_T = 0)$,

$$\tan\alpha_0 = \frac{9.80665}{2v_0^2} \frac{x_T}{\cos^2\alpha_0} e^{k \cdot x_T / \cos\alpha_0}. \qquad (4.2.4)$$

Angle of flight at a given point (x, y) on the trajectory,

$$\tan\alpha = \tan\alpha_0 - \frac{9.80665}{v_0^2 \cdot \cos^2\alpha_0} \cdot x \cdot \left(1 + \frac{k}{2\cos\alpha_0} x\right) \cdot e^{kx/\cos\alpha_0}. \qquad (4.2.5)$$

Projectile velocity at the point (x, y) is

$$v = v_0 \frac{\cos \alpha_0}{\cos \alpha} e^{-kx/2\cos \alpha_0} [1 + 2\frac{kx}{\cos \alpha_0} + \frac{1}{2}(\frac{kx}{\cos \alpha_0})^2]^{-1/2}. \quad (4.2.6)$$

Time of flight, at the point (x, y), is

$$t = \frac{1}{v_0 \cos \alpha_0} \int_0^x e^{kx/2\cos \alpha_0} [1 + 2\frac{kx}{\cos \alpha_0} + \frac{1}{2}(\frac{kx}{\cos \alpha_0})^2]^{1/2} dx. \quad (4.2.7)$$

The definite integral on the right side of (4.2.7) can be estimated using a graphing calculator.

Using (4.2.1), for the **drop**

$$\overline{y} = y - \tan \alpha_0 \cdot x,$$

of the projectile at a point with coordinates (x, y) we can write:

$$\overline{y} = -\frac{g}{2v_0^2}(\frac{x}{\cos \alpha_0})^2 e^{kx/\cos \alpha_0}. \quad (4.2.8)$$

If the bullet is fired horizontally, departure angle $\alpha_0 = 0$, the drop that corresponds to a given range x is

$$\overline{y} = -\frac{g}{2v_0^2} x^2 e^{kx}. \quad (4.2.9)$$

The above equation is obtained substituting $\alpha_0 = 0$ in equation (4.2.8). From (4.2.9) we find that the parameter k is

$$k = \frac{1}{x} \cdot \ln(-\frac{2 \cdot v_0^2 \cdot \overline{y}}{g \cdot x^2}). \quad (4.2.10)$$

The parameter k depends on the horizontal range x.

If we consider the horizontal range at the impact point, $x = x_T$, then

$$k = \frac{1}{x_T} \cdot \ln(-\frac{2 \cdot v_0^2 \cdot \overline{y}_T}{g \cdot x_T^2}). \quad (4.2.11)$$

The exponential equation of the trajectory (4.2.1), and the set of corresponding ballistics elements, (equations: (4.2.4), (4.2.5), (4.2.6), (4.2.7), and (4.2.8)) are simple equations and contain only one parameter "k".

The exponential equation of the trajectory gives acceptable results in predicting the ballistics elements of shooting with small arms.

Comment

Based on non-rigidity and equal drop model, the value of parameter k determined using (4.2.11) is practically equal to the value of k determined using (4.2.2).

Indeed, in (4.2.11) we use the drop $\bar{y}$ that corresponds to range x_T and departure angle $\alpha_0 = 0$, while in (4.2.2) we use the drop

$$\bar{y} = y_T - x_T \tan \alpha_0. \qquad (4.2.12)$$

that corresponds to departure angle $\alpha_0 \neq 0°$.

According to the non-rigidity model, the drop is the same no matter what is the departure angle.

If we substitute $\alpha_0 = 0°$ in (4.2.2) we obtain (4.2.11).

The following example (2.1) demonstrates the fact that k in (4.2.10) and (4.2.2) is the same.

Example 2.1

Consider Lapua Scenar GB528 19.44 g. (300 gr.) bullet launched with velocity $v_0 = 830 m/s$, angle $\alpha_0 = 1.147099°$.

(a) Find the parameter "k" if the point of impact corresponds to launching angle $\alpha_0 = 1.1471°$ is ($x_T = 1500$, $y_T = 0$).

(b) Find the parameter "k" if the projectile is launched horizontally ($\alpha_0 = 0°$).

(c) Use value of k obtain in (a) to find all elements of the trajectory.

Solution

Estimating k.

- **(a) Method 1**
  Substituting in (4.2.2): $x_T = 1500$, $y_T = 0$, $\alpha_0 = 1.1471°$, the initial velocity $v_0 = 830$ and solving the obtained equation, we find that the value of parameter k is
  $$k = 4.1906 \cdot 10^{-4}. \tag{1}$$

- **(b) Method 2**
  Using the principle of rigidity we find that the bullet drop at 1500 meters is
  $$\bar{y} = -x_T \cdot \tan \alpha_0 = -1000 \cdot \tan(1.1471) = -30.035 m.$$

  Using (4.2.10) we find that the parameter k is
  $$k = \frac{1}{x} \cdot \ln(-\frac{2 \cdot v_0^2 \cdot \bar{y}}{g \cdot x^2}) = \frac{1}{1500} \ln(\frac{-2 \cdot 830^2 \cdot (-30.035)}{9.80665 \cdot (1500)^2}) = 4.1924 \times 10^{-4}.$$

  Thus, the equation of the trajectory (4.2.1) for the given Lapua bullet can be written:
  $$y = \tan \alpha_0 \cdot x - \frac{9.80665}{2 \cdot (830)^2} (\frac{x}{\cos \alpha_0})^2 e^{4.1906 \times 10^{-4} \cdot x / \cos \alpha_0}. \tag{2}$$

  where $0 \leq x \leq 1500 m$.

- **(c) Elements of Trajectory**
  The y-coordinate of bullet at horizontal range $x = 1200 m$ is
  $$y = \tan(1.1471) \cdot (1200) - \frac{9.80665}{2 \cdot (830)^2} (\frac{1200}{\cos(1.1471)})^2$$
  $$e^{4.1906 \times 10^{-4} \cdot (1200) / \cos(1.1471)} = 7.0794 m$$
  ,

Substituting in (4.2.8) the value k (formula 1), and the necessary data we find that drop at $x = 1200m$ is

$$\bar{y} = -\frac{g}{2v_0^2}(\frac{x}{\cos\alpha_0})^2 e^{kx/\cos\alpha_0} = -16.9486 \ m$$

Substituting in (4.2.5) we have:

$$\tan\alpha = \tan(1.1471) - \frac{9.80665}{(830)^2 \cdot \cos^2(1.1471)} \cdot (1200) \cdot [1 +$$

$$+ \frac{4.1906 \times 10^{-4}}{2\cos(1.1471)}(1200)] \cdot e^{4.1906 \times 10^{-4}/\cos(1.1471)} = -0.0153324$$

Hence, we find that the angle of bullet in motion at the point of the trajectory with coordinates (1200, 7.0760m) is

$$\alpha = \arctan(-0.0153324) = -0.878414°.$$

Substituting in (4.2.8), we find that the velocity of the Lapua bullet is

$$v = (830)\frac{\cos(1.1471)}{\cos(-0.878414)} e^{-4.1906 \times 10^{-4}(1200)/2\cos(1.1471)}[1 + 2\frac{(4.1906 \times 10^{-4})}{\cos(1.1471)} +$$

$$+ \frac{1}{2}(\frac{4.1906 \times 10^{-4}(1200)}{\cos(1.1471)})^2]^{-1/2} = 442.06 m/s.$$

Integrating (4.2.9), for example using a graphing calculator, or any mathematics PC software (Maple, Mathematica, etc,) we find that the time of flight is $t = 2.0476$ seconds.

Note

In the same way, as in example 2.1, we find all the elements of the trajectory at the points with abscissa 300, 600, ..., 1500.

The obtained results are presented in table 2.1 below.

For comparison, in table 2.2 are given the data obtained by Doppler radar measurements:

Table 2.1 0.338 GB528 Scenar bullet. Predicted Data Using Exponential Equation

Range (m)	0	300	600	900	1,200	1,500
Velocity (m/s)	830	695	591	509	442	387
Time (s)	0.000	0.396	0.866	1.414	2.048	2.774
Drop (m)	0.000	-0.727	-3.296	-8.409	-16.949	-30.035

Table 2.2 0.338 GB528 Lapua Scenar 19.44 g bullet. Doppler radar measurements [21]

Range (m)	0	300	600	900	1,200	1,500
Velocity (m/s)	830	711	604	507	422	349
Time (s)	0.000	0.3918	0.8507	1.3937	2.0435	2.8276
Drop (m)	0.000	-0.715	-3.203	-8.146	-16.571	-30.035

Comment

Comparing table 2.1 and table 2.2, we see that the accuracy in determining the elements of trajectory of the Lapua Scenar GB528 19.44 g (300 gr) bullet for ranges until 1500 meters is acceptable (except velocity).

The largest difference in projectile drop (arround 0.38 meters) is at range 1,200 metters.

There are relatively large discrepancies in the estimation of the terminal velocity of the given projectile.

The approximate exponential formula (4.2.1) is easy to be used for any long range shooting since it requires only the launching angle (or drop) that can be obtained experimentally, or from reliabale range tables.

Example 2.2

Consider Lapua bullet of example 2.1

(a) Find the launching angle needed to hit the target located at the horizontal range 1200 meters.

(b) Find as well the coordinates of the maximum height of the trajectory.

(c) Find the launching angle that needed to zero the rifle at 1200 meters if the drop of the given projectile at 1,200 meters is 16.571m and not 16.949.

Solution

Substituting in equation (2) of example 2.1, $y = 0$, and $x = 1200$, we can write:

$$\tan\alpha_0 = \frac{9.80665}{2\cdot(830)^2}\frac{(1200)}{\cos^2\alpha_0}e^{4.18889\times 10^{-4}\cdot(1200)/\cos\alpha_0}.$$

Hence, we obtain the following equation:

$$\tan\alpha_0 = \frac{0.008541}{\cos^2\alpha_0}e^{0.50267/\cos\alpha_0}. \tag{1}$$

(a) To find the launching angle we have to solve the transcendental equation (1).

Method 1

The equation (1) can be solved for example using a graphing calculator. Thus, the solution obtained using TI84 Plus is.
$$\alpha_0 = 0.80913°.$$

Method 2

An alternative way is as follows.

Since the launching angle is small, as a first approach, we can consider $\cos\alpha_0 \approx 1$. Substituting $\cos\alpha_0 \approx 1$ in (1), we obtain:

$$\tan\alpha_0 = 0.008541e^{0.50267} = 0.0141196.$$

Hence,
$$\alpha_0 = \arctan(0.0141196) = 0.80894°. \tag{2}$$

Improving accuracy

Substituting (2) in (1) we have

$$\tan \alpha_0 = \frac{0.008541}{\cos(0.80894)} e^{0.50267/\cos(0.80894)} = 0.014123. \tag{3}$$

Thus, the departure angle is

$$\alpha_0 = \arctan(0.0141215) = 0.80914°. \tag{4}$$

We can still proceed, in the same way, to find another approximate estimation.

Method 3

Using the principle of non-rigidity of trajectory, considering the calculated drop (example 5.1), we find that the departure angle is

$$\alpha_0 = \tan^{-1}(\frac{\overline{y}_T}{x_T}) = \tan^{-1}(\frac{16.949}{1200}) = 0.80920°. \tag{5}$$

Note that the last method is the simplest one since it requires to calculate the drop using the equation (4.2.8) (as in example 2.1), and then to use the principle of non-rigidity of the trajectory.

Using the above estimated angle $\alpha_0 = 0.80920°$ and formula (4.2.1), we can easily verify that the projectile launched with the angle $\alpha_0 = 0.80920°$ will hit the target not at the center, but 0.38 meters above it, i.e. at the point with coordinates (1200m, 0.38m).

Indeed, using (4.2.1), we find:

$$y = \tan \alpha_0 \cdot x - \frac{9.80665}{2 \cdot (830)^2} (\frac{1200}{\cos 0.80920})^2 e^{4.1889 \times 10^{-4} \cdot (1200)/\cos(0.80920)}$$

$$= 0.38m.$$

(b) Maximum Height

At the maximum height of the trajectory, (x_m, y_m), the angle of projectile flight is $\alpha = 0$. Substituting in (1.7) we have:

$$\tan \alpha_0 - \frac{9.80665}{v_0^2 \cdot \cos^2 \alpha_0} \cdot x_m \cdot (1 + \frac{k}{2\cos \alpha_0} x_m) \cdot e^{kx_m/\cos \alpha_0} = 0. \quad (6)$$

The abscissa x_m of the trajectory vertex is obtained solving equation (6), after substituting $k = 4.18889 \times 10^{-4}$, $\alpha_0 = 0.80914°$, $v_0 = 830$.

We can write:

$$0.014123 - 1.423807 \cdot 10^{-5} \cdot x_m \cdot (1 + 2.094654 \cdot 10^{-4} x_m) \cdot e^{4.189308 \cdot 10^{-4} x_m} = 0. \quad (7)$$

The equation (7) can be solved using a graphing calculator, or any mathematics software, or using the trial and error procedure. The solution of equation (7) is $x_m = 842.74$.

Substituting the above value in (4.2.1), we find that the maximum height of the trajectory is $y_m = 7.195m$.

The coordinates of the trajectory vertex are ($x_m = 842.74m$, $y_m = 7.195m$).

(c) Since the real value of the drop at 1200 meters is 16.571m and not 16.949m, we find that the corresponding launching angle is

$$\alpha_0 = \tan^{-1}(\frac{\overline{y}_T}{x_T}) = \tan^{-1}(\frac{16.571}{1200}) = 0.79116°.$$

Example 2.3 For Lapua bullet of example 2.1, find the bullet drop at the following ranges: 1500, 1200, and 900 meters.

Solution

Substituting in (4.2.9), x = 1500 we find that the drop of projectile is

$$\overline{y} = -\frac{g}{2v_0^2} x^2 e^{kx} = -\frac{9.80665}{2 \cdot (830)^2} \cdot (1500)^2 \cdot e^{(4.1924 \times 10^{-4} \cdot (1500))} = -30.035m.$$

For the ranges 1200m, and 900 meters we find respectively $y = -16.951 m$, and $y = -8.408 \, m$.

4.3 Exponential Equation in Non-Standard Atmosphere

The exponential equation of the projectile trajectory in non-standard atmosphere is

$$y = \tan \alpha_0 \cdot x - \frac{g}{2v_0^2}(\frac{x}{\cos \alpha_0})^2 e^{k \cdot J^{-1} \cdot x/\cos \alpha_0} \qquad (4.3.1)$$

where

$$J^{-1} = \frac{p_0}{p_{0N}}\sqrt{\frac{\tau_{0N}}{\tau_0}}. \qquad (4.3.2)$$

The value of k can be easily found if we consider departure angle $\alpha_0 = 0$ (bullet launched horizontally):

$$k = \frac{1}{J^{-1}x_T} \cdot \ln(-\frac{2 \cdot v_0^2 \cdot \overline{y}_T}{g \cdot x_T^2}) \qquad (4.3.3)$$

where x_T and y_T are the coordinates of a terminal point on the trajectory.

Departure angle

$$\tan \alpha_0 = \frac{9.80065}{2v_0^2}\frac{x}{\cos^2 \alpha_0}e^{kJ^{-1}x/\cos \alpha_0}. \qquad (4.3.4)$$

Angle of flight

$$\tan \alpha = \tan \alpha_0 - \frac{9.80665}{v_0^2 \cdot \cos^2 \alpha_0} \cdot x \cdot (1 + \frac{k \cdot J^{-1}}{2 \cos \alpha_0}x) \cdot e^{kJ^{-1}x/\cos \alpha_0}. \qquad (4.3.5)$$

Velocity

$$v = v_0 \frac{\cos \alpha_0}{\cos \alpha} e^{-kJ^{-1}x/2\cos \alpha_0}[1 + 2\frac{kJ^{-1}x}{\cos \alpha_0} + \frac{1}{2}(\frac{kJ^{-1}x}{\cos \alpha_0})^2]^{-1/2}. \qquad (4.3.6)$$

Time

$$t = \frac{1}{v_0 \cos\alpha_0} \int_0^x e^{kJ^{-1}x/2\cos\alpha_0}[1 + 2\frac{kJ^{-1}x}{\cos\alpha_0}x + \frac{1}{2}(\frac{kJ^{-1}x}{\cos\alpha_0})^2]^{1/2} dx. \quad (4.3.7)$$

Projectile drop when departure angle is α_0 (including $\alpha_0 = 0$)

$$\bar{y} = y - \tan\alpha_0 \cdot x = -\frac{g}{2v_0^2}(\frac{x}{\cos\alpha_0})^2 e^{k \cdot J^{-1} \cdot x/\cos\alpha_0}. \quad (4.3.8)$$

Note that in standard atmosphere $J^{-1} = 1$.

The bullet drop when the departure angle is zero is

$$\bar{y} = -\frac{g}{2v_0^2}x^2 e^{k \cdot J^{-1} \cdot x}. \quad (4.3.9)$$

Example 3.1 Non-Standard Atmosphere

To determine the parameter k, which is present in formulas (4.3.1) - (4.3.8), were performed firing tests shooting horizontally (launching angle zero degree) on a target located at the distance 1000.

The 0.338 Lapua GB528 Scenar 19.44g bullet was fired with velocity 830m/s in a non-windy weather with temperature 25 degree Celsius, pressure 750mm Hg., humidity 70%.

The results of the tests (range, drop) are: (x = 1000m, drop = y = - 10.364m).

(a) Find k.

(b) Estimate the drop and velocity of the projectile flying in ICAO atmosphere (J^{-1} =1), at ranges 1000m, 900m, 600 and 300m.

(c) Estimate the drop and velocity of the projectile at ranges 1000m, 900m, 600m and 300m when shooting is done in temperature 0.00 degree Celsius, pressure 750mm Hg, humidity 50%.

Solution

To the humidity 70% and temperature 25 degree Celsius the pressure of water vapors is $e = 16.632 mm\ Hg$. (table 2.1 section 3.2).

The corresponding virtual temperature is

$$\tau_0 = \frac{T_0}{1-0.3785 \cdot e_0 / p_0} = \frac{273.15+25}{1-0.3785 \cdot 16.632/750} = 298.40°K. \quad (1)$$

The value of J^{-1} is

$$J^{-1} = \frac{p_0}{p_{0N}} \sqrt{\frac{\tau_{0N}}{\tau_0}} = \frac{750}{760}\sqrt{\frac{288.15}{298.40}} = 0.9699.$$

Substituting $x = 1000$, $y = -10.364$, departure angle $\alpha_0 = 0$ degree, and all the other necessary data in (4.3.1) we can write:

$$-\frac{9.80665}{2(830^2)}(\frac{1000}{\cos 0})^2 e^{k(0.9699)\cdot(1000)/\cos 0} = -10.364,$$

or

$$e^{969.932 \cdot k} = 1.4561057.$$

Hence,

$$k = \frac{\ln(1.4561057)}{969.932} = 3.87414 \cdot 10^{-4}. \quad (2)$$

The value of $k = 3.87414 \cdot 10^{-4}$ is valid for horizontal ranges $0 \le x \le 1000m$.

(b) Substituting $J^{-1} = 1$, $k = 3.87315 \cdot 10^{-4}$, $\alpha_0 = 0$ and all the necessary data into (4.3.1), (4.3.5), and (4.3.6) we find:

When $x = 1000m$: Drop $\bar{y} = -10.48m$, velocity v $503m/s$.
When $x = 900m$: Drop $\bar{y} = -8.17m$, velocity $v = 526m/s$.
When $x = 600m$: Drop $\bar{y} = -3.23m$, velocity $v = 605m/s$.
When $x = 300$: Drop $\bar{y} = -0.72m$, velocity $v = 704m/s$.

(c) When the humidity is 50%, the temperature is 0 degree Celsius, the pressure of water vapors is $e = 2.29mm\ Hg$.. So, we have $J^{-1} = 1.013$.

Substituting $J^{-1} = 1.013$, the value $k = 3.87315 \cdot 10^{-4}$, the departure angle equal to zero and all necessary data into (4.3.1) or (4.3.5), and in (4.3.6) we find:

For x = 1000m: Drop $\bar{y} = -10.54m$, velocity $v = 500m/s$.

For x = 900m: Drop $\bar{y} = -8.21m$, velocity $v = 523m/s$.

For x = 600m: Drop $\bar{y} = -3.24m$, velocity $v = 603m/s$.

For x = 300: Drop $\bar{y} = -0.72m$, velocity $v = 702m/s$.

4.4 Modified Exponential Equation of Trajectory

The exponential equation of bullet trajectory, shown in section 4.3, gives approximate results.

Though the differences between GB528 Lapua bullet drops, estimated using exponential equation (4.3.1) and the respective drops, measured by Doppler radar, are relatively small, the discrepancies in bullet velocities (as well as discrepancies in times) are relatively large.

To reduce those discrepancies, we introduce a correction factor "b" in equation (4.3.1).

Departure Angle Different from Zero

Introducing b, the exponential equation of the projectile trajectory in non-standard atmosphere can be written:

$$y = \tan\alpha_0 \cdot x - \frac{g}{2v_0^2}\left(\frac{x}{\cos\alpha_0}\right)^2 \cdot b \cdot e^{J^{-1}kx/\cos\alpha_0}. \qquad (4.4.1)$$

The correction factor "b" allow us to determine the equation of the trajectory of a bullet using two experimental data (or two data from the range table), i.e. the coordinates of two impact points that correspond to two different ranges.

The other equations related to the equation (4.4.1) are:

$$\tan\alpha_0 = \frac{9.80065}{2v_0^2} \frac{b \cdot x}{\cos^2\alpha_0} e^{kJ^{-1}x/\cos\alpha_0}. \qquad (4.4.2)$$

The equation (4.4.2) can be written in following form

$$\sin(2\alpha_0) = \frac{9.80065}{v_0^2} b \cdot x \cdot e^{kJ^{-1}x/\cos\alpha_0} \qquad (4.4.3)$$

Angle of flight

$$\tan\alpha = \tan\alpha_0 - \frac{9.80665}{v_0^2 \cdot \cos^2\alpha_0} \cdot b \cdot x \cdot (1 + \frac{k \cdot J^{-1}}{2\cos\alpha_0} x) \cdot e^{kJ^{-1}x/\cos}. \qquad (4.4.4)$$

Velocity

$$v = v_0 \frac{\cos\alpha_0}{\cos\alpha} \cdot b^{-1/2} e^{-kJ^{-1}x/2\cos\alpha_0} [1 + 2\frac{kJ^{-1}x}{\cos\alpha_0} + \frac{1}{2}(\frac{kJ^{-1}x}{\cos\alpha_0})^2]^{-1/2}. \qquad (4.4.5)$$

Time

$$t = \frac{b^{1/2}}{v_0 \cos\alpha_0} \int_0^x e^{kJ^{-1}x/2\cos\alpha_0} [1 + 2\frac{kJ^{-1}x}{\cos\alpha_0} + \frac{1}{2}(\frac{kJ^{-1}x}{\cos\alpha_0})^2]^{1/2} dx. \qquad (4.4.6)$$

Projectile drop

$$\bar{y} = \tan\alpha_0 \cdot x - y = \frac{g}{2v_0^2}(\frac{x}{\cos\alpha_0})^2 b \cdot e^{k \cdot J^{-1} \cdot x/\cos\alpha_0}. \qquad (4.4.7)$$

Departure Angle Equal to Zero

If the projectile is launched horizontally, departure angle $\alpha_0 = 0$, the projectile drop is

$$\bar{y} = -\frac{g}{2v_0^2} \cdot b \cdot x^2 e^{J^{-1}kx}. \qquad (4.4.8)$$

The other equations related with (4.4.8) can be obtained substituting $\alpha_0 = 0$ in (4.4.2) - (4.4.7). Thus we have:

Modified Exponential Equation of the Trajectory

$$y = -\frac{g}{2v_0^2} b \cdot x^2 \cdot e^{J^{-1}kx/\cos\alpha_0}. \qquad (4.4.9)$$

which, at the same time, is the projectile drop (4.4.8).

Angle of flight

$$\tan\alpha = -\frac{9.80665}{v_0^2} \cdot b \cdot x \cdot (1 + \frac{k \cdot J^{-1}}{2} x) \cdot e^{kJ^{-1}x}. \qquad (4.4.10)$$

Velocity

$$v = \frac{v_0}{\cos\alpha} \cdot b^{-1/2} e^{-kJ^{-1}x/2} [1 + 2k \cdot J^{-1}x + \frac{1}{2}(k \cdot J^{-1}x)^2]^{-1/2}. \qquad (4.4.11)$$

Time

$$t = \frac{b^{1/2}}{v_0} \int_0^x e^{kJ^{-1}x/2} [1 + 2kJ^{-1}x + \frac{1}{2}(kJ^{-1}x)^2]^{1/2} dx. \qquad (4.4.12)$$

Note that in standard atmosphere $J^{-1} = 1$.

Example 4.1

ICAO Atmosphere ($J^{-1} = 1$), departure velocity standard, $v_0 = 830 m/s$.

Determine the parameters "b" and "k" using the data of the GB528 Lapua Scenar bullet shown in table 2.2, section 4.2, i.e.

Departure angle, $\alpha_0 = 0$.

Range $x_1 = 1200$ meters, drop $\overline{y}_1 = -16.71$ m,

Range $x_2 = 1500$ meters, drop $\overline{y}_2 = -30.035$ m. (1)

Solution

Using equation (4.4.8) for two ranges we can write

$$\frac{\bar{y}_2}{\bar{y}_1} = (\frac{x_2}{x_1})^2 e^{k(x_2-x_1)}. \tag{2}$$

Substituting the data given above in (1), and solving for k, we find:

$$k = 4.94740287 \times 10^{-4}. \tag{3}$$

Substituting in (4.4.8), $x_2 = 1500$, $\bar{y}_2 = -30.035$ m, and k given in (3), and then solving for "b", we find.

$$b = 0.89292742. \tag{4}$$

The table 4.1 is obtained using the equations (4.4.2) - (4.4.8), and the parameters (3) and (4).

Table 4.1 GB528 bullet. Predicted Data Using Modified Exponential Equation

Range (m)	0	300	600	900	1,200	1,500
Velocity (m/s)	830	713	592	499	425	365
Time (s)	0.000	0.396	0.914	1.581	2.430	3.497
Drop (m)	0.000	-0.664	-3.079	-8.036	-16.571	-30.035

Note

Comparing the predicted data of table 4.1 with Doppler radar data (table 2.2 section 4.2), we see that the modified exponential equation of the trajectory gives improved approximation data, close to Doppler radar data.

Example 4.2 Inclined Shooting

Use the modified equation of trajectory related with GB528 Lapua bullet, with $k = 4.94740287 \times 10^{-4}$ and $b = 0.89292742$, found in example 4.1, to find the departure angle and the super elevation angle needed to hit a target located at the inclined range $D = 1200m$, if the elevation angle is $E = 30°$.

The firearm is on the ground with temperature 0.00 degree Celsius, pressure 750mm Hg, humidity 50%.Consider the temperature of cartridge 21.11 degree Celsius.

The value of J^{-1}, calculated in example is 3.1 is

$$J^{-1} = \frac{p_0}{p_{0N}}\sqrt{\frac{\tau_{0N}}{\tau_0}} = \frac{750}{760}\sqrt{\frac{288.15}{298.40}} = 0.9699.$$

Solution

The coordinates of the target are

$$x_T = D \cdot (\cos(E)) = 1200 \cdot \cos(30) = 1039.23 m$$

and

$$y_T = D \cdot \sin(E) = 1200 \cdot \sin(30) = 600 m$$

Substituting in equation (4.4.1),

$$y = \tan\alpha_0 \cdot x - \frac{g}{2v_0^2}(\frac{x}{\cos\alpha_0})^2 \cdot b \cdot e^{J^{-1}kx/\cos\alpha_0}$$

we have

$$600 = \tan\alpha_0 \cdot (1039.23) - \frac{g}{2(830)^2}(\frac{1039.23}{\cos\alpha_0})^2 \cdot (0.89292742) \cdot$$

$$\cdot e^{(0.9699)\cdot(4.94703\times10^{-4})\cdot(1039.23)/\cos\alpha_0}$$

Solving the above equation with a graphing calculator we find the departure angle

$$\alpha_0 = 30.68052°.$$

The super elevation angle is

$$\alpha_{0i} = 0.68052°.$$

Note: Using equation (2.2.1), we find the approximate value

$$\alpha_{OI} = \alpha_0 \cdot \cos(E) = 0.79116 \cdot \cos(30) = 0.6852°.$$

There is an irrelevant difference between two values of 0.27 minutes.

4.5 Effect of Cartridge Temperature on Projectile Trajectory

The elements of projectile trajectory, presented in range tables of a firearm, are calculated for the standard atmospheric conditions and for some standard ballistics characteristics of the projectile, including the standard departure velocity.

In all approximate equations of projectile trajectories in non standard atmosphere we have considered in preceding sections, the initial velocity of projectile is equal to the standard initial velocity v_0.

So, regardless of the temperature of air at shooting site, the initial velocity is equal to the initial velocity in standard atmosphere.

That is true if we are able to store the cartridges in a temperature that is equal to the temperature of dry air T_0 in standard atmosphere ($T_0 = 21.11$ degree Celsius in ICAIO and ASM, $T_0 = 15$ degree Celsius in TSA).

The temperature of cartridge (propellant charge) must be kept standard in experimental shootings.

The meteorological characteristics of the atmosphere at a shooting site and the projectile ballistic characteristics are usually different from the standard ones.

In general, the marksman has no possibility to store the cartridges in standard temperature T_{0N}.

Temperature of propellant charge is equal to the temperature T at the firing site which in general is different from the standard temperature T_{0N}.

The initial velocity V_0 of a bullet in non standard atmosphere can be estimated approximately by the formula[22]

$$V_0 = v_0 \cdot [1 + 0.001 \cdot (T - T_{0N})], \qquad (4.5.1)$$

where T_{0N} and v_0 are respectively the temperature of dry air and the standard initial velocity of projectile.

Note that the coefficient (0.001) in (4.5.1) expresses the efficiency of black powder. The efficiency of black powder changes from one type to another.

For some rifle powders instead of 0.001 we can use 0.0014, which is obtained using the data given in Ballistica22.[23]

For IMR powders (IMR, Improved Military Rifles), using the data provided by Rinker[24], we conclude that the coefficient of efficiency is 0.001766.

The change of initial velocity, as result of deviation of temperature from the standard one, changes the projectile standard trajectory.

To predict the projectile trajectory, when the initial velocity V_0 is not equal to standard (v_0), we consider that for small changes in initial velocity of projectile the equations obtained in sections (4.3), (4.7) and, (4.8) are still valid when we substitute velocity v_0 with V_0.

The following examples illustrate the prediction of trajectory using equation (4.5.1).

Example 5.1 Non standard initial velocity

Consider Exercise 3.1, question c.
The temperature of air at shooting site is zero degree Celsius.
The temperature of bullet at the shooting site is equal to zero degree Celsius.
As result the launching velocity of the bullet is
$$V_0 = v_0(1+0.001(t-21.11)) = 830 \cdot (1+0.001 \cdot (0.0-21.11)) = 812.49 \text{ m/s}.$$

We have (see example 3.1):
$$J^{-1} = 1.013, \ k = 3.87315 \cdot 10^{-4}, \ x = 1000.$$

Substituting the above values in (4.3.9), we find that the drop at 1000 meters is
$$\overline{y} = -\frac{g}{2v_0^2}x^2 e^{k \cdot J^{-1} \cdot x} = -\frac{9.80665}{2 \cdot (812.49)^2}(1000)^2 e^{3.187315 \times 10^{-4} \cdot (1.013) \cdot (1000)}$$

$$= -10.99m$$

Note that the value of projectile drop predicted using PC program RLAPUA16.BAS is -10.987.

Example 5.2 Non-Standard Atmosphere

To determine the parameters b and k of the modified exponential trajectory (4.4.1) there were done some firing tests by shooting horizontally (launching angle zero degree) on two targets respectively at 1000m and 700m.

The 0.338 Lapua GB528 Scenar 19.44g. bullet was fired with velocity 830m/s in a non-windy weather with temperature 25 degree Celsius, pressure 758mm Hg., humidity 70%.

The results of shooting tests (range, drop) are:
($x = 1000$ m, drop $\bar{y} = -10.208$ m), ($x = 700$ m, $\bar{y} = -4.419$ m).

(a) Find the parameters b and k presented in modified exponential trajectory.

(b) Use the obtained formula to estimate the drop of the projectile flying in the ICAO atmosphere ($J^{-1} = 1$), at ranges 1000m, 900m and 600m.

(c) Find the departure angle needed to zero the rifle at 1000 meters if the shooting is in ICAO atmosphere.

(d) Use the obtained equation of modified exponential trajectory to estimate the drop of projectile at 1000 meters when shooting is done in temperature 0.00 degree Celsius, pressure 750mm, humidity 50%.

(e) Find as well the drop and velocity of projectile at horizontal range 900 meter.

The ballistics characteristics (mass and diameter of bullet) are standard, temperature of black powder 25 degree Celsius.

Solution

(a) Using (3.2.7), we find that the pressure of water vapors is

$$e_{70\%} = 0.70 \cdot (7.50187 \cdot e^{19.04 \cdot (1. - 280.07/298.15)}) = 16.66 mm\ Hg.$$

The corresponding virtual temperature is

$$\tau_0 = \frac{T_0}{1 - 0.3785 \cdot e_0 / p_0} = \frac{273.15 + 25}{1 - 0.3785 \cdot 16.661 / 758} = 300.65°K. \quad (1)$$

The value of scaling factor J^{-1} is

$$J^{-1} = \frac{p_0}{p_{0N}}\sqrt{\frac{\tau_{0N}}{\tau_0}} = \frac{758}{760}\sqrt{\frac{288.15}{300.65}} = 0.9764. \quad (2)$$

The initial velocity of bullet is

$$V_0 = 830 \cdot (1 + 0.001 \cdot (25 - 21.11)) = 833.22 m/s.$$

Substituting in (4.4.8)

$$\alpha_0 = 0,\ v_0 = 833.22,\ x = 1000,\ \bar{y} = -10.208,\ g = 9.80665,$$

we have

$$-10.208 = -\frac{9.80665}{2 \cdot (833.22)^2} \cdot (1000)^2 \cdot b \cdot e^{(0.9764) \cdot (1000) k} \quad (3)$$

In the same way, substituting in (4.4.8),

$$\alpha_0 = 0,\ v_0 = 833.22,\ x = 700,\ \bar{y} = -4.419,\ g = 9.80665,$$

we obtain the following equation:

$$-4.419 = -\frac{9.80665}{2 \cdot (833.22)^2} \cdot (700)^2 \cdot b \cdot e^{(0.9764) \cdot (700) k}. \quad (4)$$

Dividing (3) and (4) we have

$$2.3100 = (1.42857)^2 \cdot e^{(0.9764) \cdot (300) k}. \quad (5)$$

Solving (5) we find that

$$k = 4.22974 \times 10^{-4}. \quad (6)$$

Substituting (6) in (3) we have

$$-10.208 = -\frac{9.80665}{2 \cdot (833.22)^2} \cdot (1000)^2 \cdot b \cdot e^{(0.9764) \cdot (1000) \cdot (0.000422974)}.$$

Solving, we find

$$b = 0.9563333.$$

Thus, the modified exponential equation of trajectory (4.7.5) is

$$y = \tan\alpha_0 \cdot x - \frac{g}{2(v_0)^2}(\frac{x}{\cos\alpha_0})^2 \cdot (0.9563333) \cdot e^{4.22974 \times 10^{-4} \cdot J^{-1} x / \cos\alpha_0}, \quad (7)$$

It is valid for $x \leq 1000m$.

(b) Substituting in (7): $J^{-1}=1$, $\alpha_0 = 0$, $x = 1000$, we find that at $x = 1000$ meters the drop in ICAO atmosphere is

$$\bar{y} = -\frac{9.80665}{2(830)^2}(1,000)^2 \cdot (0.9563333) \cdot e^{4.22974 \times 10^{-4} \cdot (1) \cdot (1000)} = -10.391m.$$

In the same way we find:

Range x = 900m, drop is $\bar{y} = -8.068$;

Range x = 600m, drop $\bar{y} = -3.158$.

(c) Using the non-rigidity principle of bullet trajectory we find that at 1000 meters the departure angle is

$$\alpha_0 = \frac{|\bar{y}|}{x_T} \cdot \frac{180}{\pi} = \frac{10.391}{1000} \cdot \frac{180}{\pi} = 0.5954°.$$

(d) When the humidity is 50%, the temperature is 0 degree Celsius then the pressure of water vapors is $e = 2.29mm\ Hg$.

Bullet velocity is

$$V_0 = 830 \cdot (1 + 0.001 \cdot (0.00 - 21.11)) = 812.48m/s$$

The virtual temperature is

$$\tau_0 = \frac{T_0}{1 - 0.3785 \cdot e_0/p_0} = \frac{273.15 + 0}{1 - 0.3785 \cdot 2.29/750} = 273.47°K,$$

while

$$J^{-1} = \frac{p_0}{p_{0N}}\sqrt{\frac{\tau_{0N}}{\tau_0}} = \frac{750}{760}\sqrt{\frac{288.15}{273.47}} = 1.0130.$$

Substituting in (7) the scaling factor $J^{-1} = 1.0130$ and velocity $V_0 = 812.48m/s$ $\alpha_0 = 0$ we find that the drop at 1,000 meters is

$$\bar{y} = -\frac{g}{2(812.48)^2}(1000)^2 \cdot (0.9563333) \cdot e^{4.22974 \times 10^{-4} \cdot (1.013) \cdot (1000)} = -10.90m. \quad (8)$$

Note that the drop at 1,000 meters obtained using PC program RLAPUA16.BAS is $\bar{y} = -10.987m$.

Note as well that there is a small difference of about 0.09 meters.

(e) The drop at 900 meters is

$$\bar{y} = -\frac{g}{2(812.48)^2}(900)^2 \cdot (0.9563333) \cdot e^{4.22974 \times 10^{-4} \cdot (1.013) \cdot (900)} = -8.46m$$

Let's find the angle of flight α at 900 meter. Substituting in (4.4.10) we have:

$$\tan \alpha = -\frac{9.80665}{(812.48)^2} \cdot (0.956333) \cdot 900 \times$$

$$\times [1 + \frac{(4.22974) \cdot 10^{-4} \cdot 1.013}{2}(900)] \cdot e^{(4.22974 \times 10^{-4} (1.013) \cdot (900))} = -0.02243$$

Hence,

$$\alpha = \arctan(-0.02243) = -1.2848°.$$

Velocity

Substituting in (4.4.11) we find that

$$v = \frac{812.48}{\cos(-1.2848)} \cdot (0.956333)^{-1/2} e^{-(4.22974 \times 10^{-4} (1.013) \cdot (900)/2)} \times$$

$$\times [1 + 2 \cdot (4.22974 \cdot \times 10^{-4} (1.013) \cdot 900 + \frac{1}{2}(4.22974 \cdot 10^{-4} \cdot 1.013 \cdot 900)^2]^{-1/2}$$

$$= 504.44 m/s$$

$$v = \frac{v_0}{\cos \alpha} \cdot b^{-1/2} e^{-kJ^{-1}x/2}[1 + 2k \cdot J^{-1}x + \frac{1}{2}(k \cdot J^{-1}x)^2]^{-1/2}$$

4.6 Approximate Equations to Estimate the Coriolis Effect

To find the Coriolis influence on the projectile trajectory we should solve numerically the system of equation (5.5.4).

In general, for long range shooting until 1400 meters, the cross deflection is around a maximum of 0.18 meter at the latitude 45 degree North (Azimuth 0 degree or 90 degree East).[25]

The vertical deviation from the center of target has more or less the same maximum value.

The correction for Coriolis Effect can be ignored in favor of the practicality of shooting, especially for ranges around 1000 meter.

To have an idea about the Coriolis Effect in long range shooting we will show the approximate method presented by Shapiro in his wonderful book Exterior Ballistics, Moscow 1946.

According to Shapiro, change in range Δx_T and the cross deflection Z can be estimated approximately using the following formulae:

$$\Delta x_T = (K \cdot \cos \Lambda \cdot \sin A) \cdot (\frac{g \cdot \Omega \cdot t^3}{6}), \qquad (4.6.1)$$

and

$$Z = (\frac{3 \cdot \sin \Lambda}{\sqrt{\tan \alpha_0 \cdot |\tan \alpha_T|}} - \cos \Lambda \cdot \cos A) \cdot \frac{g \cdot \Omega \cdot t^3}{6}, \qquad (4.6.2)$$

where

$$K = \frac{3 - \tan \alpha_0 \cdot |\tan \alpha_T|}{\tan \alpha_0 \cdot |\tan \alpha_T|}, \qquad (4.6.3)$$

t is the time of flight to range x_T, α_0 and α_T are respectively the departure angle and terminal angle.

For the vertical deviation of the projectile from the center of target we have:

$$\Delta y_T = -\Delta x_T \cdot \tan(\alpha_T). \qquad (4.6.4)$$

Comment

Example 6.1 (below) shows that at a given latitude in Northern hemisphere, the lateral deviation of bullet to the right due to Coriolis Effect is the same no matter what is the azimuth direction of shooting.

Example 6.1 Coriolis Effect in SI units

Find the vertical deviation and the cross deflection of 0.338 GB528 Lapua Scenar 19.44 g. bullet that results from Coriolis effect at the following range:

$x_T = 1500m$ if departure angle is $\alpha_0 = 1.141°$, terminal angle $\alpha_T = -2.0385°$, time of flight $t = 2.824s$. Angular velocity of Earth rotation is $\Omega = 7.292 \times 10^{-5} s^{-1}$.

(a) Shooting is in latitude $\Lambda = 45°$, azimuth $A = 90°$, shooting East.

(b) Shooting is in latitude $\Lambda = 45°$, azimuth $A = 0°$, Shooting North, along the meridian.

(c) Shooting is in latitude $\Lambda = 45°$, azimuth $A = 270°$, shooting West.

Solution

First we calculate
$$\frac{g \cdot \Omega \cdot t^3}{6} = \frac{9.80665 \cdot (7.292 \times 10^{-5}) \cdot (2.824)^3}{6} = 0.002684m$$

and
$$K = \frac{3 - \tan\alpha_0 \cdot |\tan\alpha_T|}{\tan\alpha_0 \cdot |\tan\alpha_T|} = \frac{3 - \tan(1.141) \cdot |\tan(-2.0385)|}{\tan(1.141) \cdot |\tan(-2.0385)|}$$

$$= 4230.842.$$

(a) Substituting in (4.6.1) and (4.6.4) and we find:

Change in range

$$\Delta x_T = (K \cdot \cos \Lambda \cdot \sin A) \cdot (\frac{g \cdot \Omega \cdot t^3}{6}) =$$

$$= (4230.84) \cdot \cos(45) \cdot \sin(90) \cdot (0.002684) = 8.030 m$$

Change in vertical direction

$$\Delta y_T = -\Delta x_T \cdot \tan(\alpha_T) = -(8.030) \cdot \tan(-2.0385) = 0.286 m.$$

The bullet will hit $| \Delta y_T |=| 0.286 |= 0.286 m$ above the center of the target.

Cross Deflection

$$Z = (\frac{3 \cdot \sin \Lambda}{\sqrt{\tan \alpha_0} \cdot | \tan \alpha_T |} - \cos \Lambda \cdot \cos A) \cdot \frac{g \cdot \Omega \cdot t^3}{6} =$$

$$= [\frac{3 \cdot \sin(45)}{\sqrt{\tan(1.141)} \cdot | \tan(-2.0385) |} - \cos(45) \cdot \cos(90)] \cdot (4230.84) = 0.214 m$$

To hit the center of the target we need to adjust sighting by shifting the aiming point 21.4 centimeter below the center and 21.4 centimeter to the left of the center.

(b) In the same way, for shooting along the meridian, (azimuth $A = 0°$), we find:

$$\Delta x_T = 0m, \ \Delta y_T = 0m.,$$

and

$$Z = (\frac{3 \cdot \sin \Lambda}{\sqrt{\tan \alpha_0} \cdot | \tan \alpha_T |} - \cos \Lambda \cdot \cos A) \cdot \frac{g \cdot \Omega \cdot t^3}{6} =$$

$$= [\frac{3 \cdot \sin(45)}{\sqrt{\tan(1.141)} \cdot | \tan(-2.0385) |} - \cos(45) \cdot \cos(0)] \cdot (0.002684) = 0.212 m$$

(c) In the same way as in (a) we find:

Change in range

$$\Delta x_T = (K \cdot \cos \Lambda \cdot \sin A) \cdot (\frac{g \cdot \Omega \cdot t^3}{6}) =$$

$$= (4230.84) \cdot \cos(45) \cdot \sin(270) \cdot (0.002684) = -8.030 m$$

Vertical deviation

$$\Delta y_T = -\Delta x_T \cdot \tan(\alpha_T) = -(-8.030) \cdot \tan(-2.0385) = -0.286 m.$$

Cross Deflection

$$Z = (\frac{3 \cdot \sin \Lambda}{\sqrt{\tan \alpha_0 \cdot |\tan \alpha_T|}} - \cos \Lambda \cdot \cos A) \cdot \frac{g \cdot \Omega \cdot t^3}{6} =$$

$$= [\frac{3 \cdot \sin(45)}{\sqrt{\tan(1.141) \cdot |\tan(-2.0385)|}} - \cos(45) \cdot \cos(270)] \cdot (4230.84) = 0.214 m$$

Note the lateral deviation of bullet to the right of shooting plane, due to Coriolis Effect is practically the same no matter what is the azimuth direction of shooting.

Example 6.2 Coriolis Effect in Imperial Units

Find the vertical deviation and the cross deflection of 0.338 GB528 Lapua Scenar 19.44 g. bullet that results from Coriolis effect at the following range:

$x_T = 1200$ yard if departure angle is $\alpha_0 = 0.6831°$, terminal angle $\alpha_T = -1.032°$, time of flight $t = 1.801 s$.

Angular velocity of Earth rotation is $\Omega = 7.292 \times 10^{-5} s^{-1}$.

Shooting is in latitude $\Lambda = 45°$, azimuth $A = 0°$ (shooting North), $g = 32.174 ft/s^2$.

Solution

First we calculate:

$$\frac{g \cdot \Omega \cdot t^3}{6} = \frac{32.174 \cdot (7.292 \times 10^{-5}) \cdot (1.801)^3}{6} = 0.002284 \, ft.$$

and

$$K = \frac{3 - \tan \alpha_0 \cdot |\tan \alpha_T|}{\tan \alpha_0 \cdot |\tan \alpha_T|} = \frac{3 - \tan(0.6381) \cdot |\tan(-1.032)|}{\tan(0.6381) \cdot |\tan(-1.032)|} = 13967.02.$$

Substituting in (4.6.1), (4.6.4) and (4.6.2) we find:

Change in range

$$\Delta x_T = (K \cdot \cos \Lambda \cdot \sin A) \cdot (\frac{g \cdot \Omega \cdot t^3}{6}) =$$

$$= (13,967.02) \cdot \cos(45) \cdot \sin(0) \cdot (0.002284) = 0 \, ft$$

Change in vertical direction

$$\Delta y_T = -\Delta x_T \cdot \tan(\alpha_T) = -(0) \cdot \tan(-1.032) = 0.$$

The bullet will hit at the center of the target.

Cross Deflection

$$Z = (\frac{3 \cdot \sin \Lambda}{\sqrt{\tan \alpha_0 \cdot |\tan \alpha_T|}} - \cos \Lambda \cdot \cos A) \cdot \frac{g \cdot \Omega \cdot t^3}{6} =$$

$$= [\frac{3 \cdot \sin(45)}{\sqrt{\tan(0.6831) \cdot |\tan(-1.032)|}} - \cos(45) \cdot \cos(0)] \cdot (0.002284) = 0.329 \, ft = 3.95 \, in$$

Thus, there is no change in range and no vertical shift over the bull's eye.

In the z-direction the deflection is 3.95 inches.

To hit the center of the target we need to adjust aiming by shifting it 3.95 inch to the left of the target center.

4.7 Determination of Coriolis Effect, Spin Drift and Wind Effect

As shown in section 4.6 the Coriolis Effect modifies the point mass projectile trajectory in long range shooting.

In general, there is a vertical deviation of bullet from the center of target and almost a constant lateral right-side deviation for shooting in North hemisphere in whatever azimuth shooting direction.

When shooting along meridian there is no vertical deviation of the projectile from the center of target.

The rotation of a projectile modifies as well the projectile trajectory, mainly diverting the impact point from the center of target to the right of launching plane xoy, when the projectile spin is right-handed, otherwise to the left.

The entire trajectory is curved to the right of the shooting plane xoy.

To eliminate the vertical deviation on the shooting plane, due to Coriolis Effect, the preferred azimuth shooting direction is along meridian.

Experimental Determination of Coriolis Deflection

Using that fact presented in the comment in section (4.6), to measure in practice of shooting the Coriolis deflection we can use the McCoy's idea (method), presented in Modern Exterior Ballistics, page 181.

McCoy's method of shooting on a given range, once in a certain azimuth direction and after (with the same firearm) in opposite direction, does not eliminate the gyroscopic drift always to the right of shooting plane for right-hand twist of barrel grove.

To eliminate the spin drift we can use the following approach.

- First, we shoot on bull's eye, at a given range, let's say 1500 meters, in East direction.

The total deviation Z_1 to the right of shooting plane xoy is

$$Z_1 = Z_C + Z_S + Z_W, \qquad (4.7.1)$$

where Z_C, Z_S and Z_W are respectively the Coriolis deviation (to the right of target), the spin drift (to the right of target) and the wind drift (assumed to the right.

- Second, with another identical rifle, but with barrel grove twist to the left, we fire in opposite direction at the same range.

  The total deviation Z_2 of the actual shooting plane xoy is
  $$Z_2 = Z_C - Z_S - Z_W \qquad (4.7.2)$$

  since Z_S and Z_W are directed to the left of plane of shooting.

The sum of (4.7.1) and (4.7.2) is

$$Z_1 + Z_2 = 2Z_C \qquad (4.7.3)$$

From (4.7.3) we find that the lateral deviation due to Coriolis effect is

$$Z_C = \frac{Z_1 + Z_2}{2} \qquad (4.7.4)$$

Note that this method eliminates as well the influence of cross wind. The shooting tests can be done as well in absence of wind.
A better result we will have if both rifles fires at the same time.

Experimental Determination of Coriolis Vertical Deviation

In a non-windy weather, using the vertical deviation of projectile when we measure the lateral Coriolis deviation, shooting in two opposite directions, we can find that the vertical deviation.

Indeed, in the above shooting experiment, measuring the vertical deviation Y_1, when shooting East and then the vertical deviation Y_2 when shooting West, we find that the absolute value of the vertical deviation is

$$Y_C = \pm \frac{Y_1 + (-Y_2)}{2}, \qquad (4.7.5)$$

where " + " sign is for shooting East, "-" for shooting West.

Experimental Determination of Spin Drift

Following Shapiro's method (idea) we can measure the spin drift by eliminating the Coriolis lateral deviation.

With both rifles mentioned above we fire at a given range (let's say 1500 meters) at the same time on two different bull's eyes.

The shooter, with the right-hand twisted barrel grove is to the right, the other shooter to the left of the first shooter. Assume that the cross-wind blows to the right.

- The deviation of bullet, fired from the right-hand spinning rifle, from bulls eye is

$$Z_1 = Z_C + Z_S + Z_W. \tag{4.7.6}$$

- The deviation of bullet, fired from the left-hand rifle, from the bull's eye is

$$Z_2 = Z_C - Z_S + Z_W. \tag{4.7.7}$$

- The difference of (4.5.5) and (4.5.6) is

$$Z_1 - Z_2 = 2Z_S. \tag{4.7.8}$$

From (4.7.8) we find that the lateral deviation due to spinning projectile is

$$Z_S = \frac{Z_1 - Z_2}{2}. \tag{4.7.9}$$

Note

To find an equation that describes one of the effects, presented in this section, it is necessary to collect shooting test data at least for 5-6 ranges, starting from around 700 meters, or 800 meters.

Then, using regression methods for each effect, we can find corresponding equation, for example, an equation that relates the spin drift with shooting range for a particular rifle and bullet in use.

ELEMENTS OF EXTERIOR BALLISTICS | 145

4.8 Range-Wind and Cross-Wind

An important factor that modifies the projectile motion is wind. Wind is a very complicated motion of an enormous mass of turbulent air.

The velocity of wind changes in magnitude and direction, and depends on the location and time the velocity is measured, as well as on the altitude and the characteristics of the shooting site (field, forest, valley, town, hills, mountains, city etc.).

The trajectory of a projectile flight in a windy weather is different from the trajectory of the same projectile in absence of wind.

Wind deflects the trajectory of the projectile in the direction of motion (x-axis) and in the perpendicular direction (z-axis). As result, the projectile will miss the center of target if the launching angle is set up to hit the target in absence of wind.

Wind is characterized by the velocity-vector $\vec{w}$ that can be seen as composed by the "range wind" $\vec{w}_x$, which blows in the direction of fire, or in opposite direction of fire (along x-axis), and the "cross wind" $\vec{w}_z$ that blows perpendicular to the shooting plane.

Thus,

$$\vec{w} = \vec{w}_x + \vec{w}_z .\tag{4.8.1}$$

The component of wind in vertical direction (y-axis) is not considered.

In general, the velocity of wind (4.8.1) can change from one shooting to another.

To predict the projectile trajectory in presence of wind, the velocity of wind (4.8.1) is considered constant during the entire trajectory and equal to an average value (expected value).

Projectile Motion in Presence of Range Wind

The range-wind component of a uniform wind that blows with constant velocity $\vec{w}_x$ changes the velocity of the projectile and the angle of motion. Range wind (as well as cross-wind), acts on projectile during entire trajectory.

Quite immediately, after the projectile leaves the muzzle of firearm, the departure velocity and the departure angle change.

The constant range-wind that blows in the direction of shooting, increases the component of the horizontal velocity of projectile, and decreases the departure angle.

The range wind that blows in the opposite direction of shooting, decreases the initial velocity of projectile, and increases the launching angle.

Because of the range-wind, $\vec{w}_x$, the initial velocity of a projectile, quite instantly after it leaves the muzzle of firearm, becomes

$$\vec{v}_{w0} = \vec{v}_0 + \vec{w}_x. \qquad (4.8.2)$$

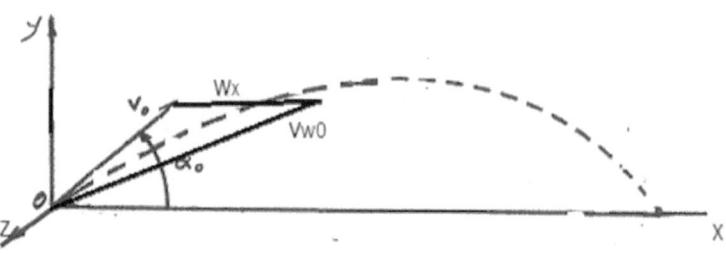

"Figure 7"

Experimental Determination of Range-Wind Vertical Deviation

- Shooting along the meridian in a windy weather we can measure the vertical deviation of the projectile from the center of target using the procedure shown in **Experimental Determination of Coriolis Vertical Deviation** (section 4.7)
- Since we have measured the vertical deviation caused by Coriolis Effect (in a non-windy weather), the shooting direction is not important, but we need to exclude the contribution of already measured Coriolis Effect.

Based on experimental data it is possible to derive experimental equations to measure the range-wind vertical deviation.

Note that a theoretical method is shown in my books EBA and EBRM, where the influence of range-wind in projectile trajectory is predicted as correction.

The PC programs presented in the book take into account the range-wind and cross-wind deflection.

Experimental Determination of Cross-Wind Drift

The effect of range wind can be studied experimentally using the method presented in section 4.7 above.

In the section 4.7, we have shown a practical method to measure the Coriolis drift Z_C. We can determine the cross-wind drift Z_W, after we have measured Coriolis drift.

Indeed, adding (4.7.5) and (4.7.6) we find that

$$Z_1 + Z_2 = 2Z_C + 2Z_W. \qquad (4.8.5)$$

Hence, we find that the cross-wind drift is

$$Z_W = \frac{Z_1 + Z_2 - 2Z_C}{2}. \qquad (4.8.6)$$

The deviation of the impact point in the perpendicular direction to the departure plain of projectile, i.e. in the positive direction of the z-axis, or, in opposite direction is given by Didion's equation

$$z_{Wc} = w_c (t_T - \frac{x_T}{v_0 \cos \alpha_0}), \qquad (4.8.7)$$

where w_c is velocity of cross wind.

According to McCoy the Didion's classical formula is very accurate.

Note that the PC programs associated with all my books estimate the cross wind deflection based on Didion's formula.

Example 8.1

Consider a 338 Lapua GB528 Scenar 19.44g (300 grain) bullet fired with a departure speed of $v_0 = 830 m/s$ at the sea level in the ICAO.

The elements of the trajectory at the horizontal range of 1000 meters are:

Departure angle, $\alpha_0 = 0.5978°$, time of flight, $t = 1.592s$.

Estimate the cross-wind deflection if there is a cross wind component of $w_c = 4m/s$.

Solution

The cross-wind deflection in the positive direction of z-axis is

$$z_{wc} = w_c(t_c - \frac{x_T}{v_0 \cos \alpha_0}) = 4 \cdot (1.592 - \frac{1000}{830 \cdot \cos(0.5978)}) = 1.55m.$$

The deflection angle of the impact point is

$$\alpha_z = \arctan(1.55/1000) = 0.089°.$$

4.9 Newton-Snell's Law in Exterior Ballistics

Two "ballistic projectile trajectories" that have geometric, kinematics and dynamic similitude, at their corresponding points, have similar angles of flights and similar scaling elements (velocity, vertex, time, departure angles, and terminal angles) expressed mathematically by similarity equations:

Similarity Equations

- Corresponding coordinates (x, y),
$$x_2 = J \cdot x_1, \quad y_2 = J^2 \cdot y_1. \tag{4.9.1}$$

- Corresponding times,
$$t_2 = J \cdot t_1. \tag{4.9.2}$$

- Corresponding velocities,
$$v_2 = v_1. \tag{4.9.3}$$

- Corresponding drops
$$\bar{y}_2 = J^2 \cdot \bar{y}_1. \tag{4.9.4}$$

- Corresponding angles of flights, α_2, α_1, are related according to Newton-Snell's law

$$\sin \alpha_2 = J \sin \alpha_1, \qquad (4.9.5)$$

where $\cos \alpha_2 = \cos \alpha_1$.

Newton-Snell's law expresses the relation between corresponding angles (α_2, α_1) of projectile trajectories, at two corresponding locations (i.e. corresponding points on two respective similitude trajectories).

Scaling factor is denoted by J. Scaling factor that relates the corresponding elements of both trajectories is

$$J = (\rho_{0N} a_{0N}) / (\rho_0 a_0),$$

where ρ_{0N} and a_{0N} are respectively the density and speed of sound at a point (x_1, y_1) on the first trajectory, usually in standard atmosphere, while ρ_0 and a_0 are respectively the density and speed of sound at the corresponding point (x_2, y_2) on the second trajectory (non-standard atmosphere).

Scale factor J is the "index of refraction" that represents the ratio of acoustic impedance of air in standard atmosphere and acoustic impedance of air in non-standard atmosphere.

Though the projectile does not actually pass through a physical interface that separates two atmospheres, the equation (4.9.5) is just the Snell's law of wave refraction as it is interpreted by one of the greatest scientist Sir Isaac Newton.

Scaling factor can be written

$$J = (\frac{p_{0N}}{p_0} \sqrt{\frac{\tau_0}{\tau_{0N}}}) = (\frac{\rho_{0N}}{\rho_0} \cdot \sqrt{\frac{\tau_{0N}}{\tau_0}}) = (\frac{\rho_{0N}}{\rho_0} \frac{a_{0N}}{a_0}). \qquad (4.9.6)$$

As it is shown in EBNA, the set of similarity equations and the Newton-Snell's law equation, applied to motion of projectiles, give remarkably accurate results for the solution of problems of exterior ballistics and to predict the projectile trajectory, not only for small arms but also for the artillery fire.

The accuracy is better for relatively not large launching angles.

In general, the Newton-Snell's law and the other related equations are valid for any two atmospheres, standard or not. **In other words,** Newton-Snell's law relates two non-standard atmosphere trajectories with different acoustic impedances.

If we denote

$$J_{1,2} = (\rho_{01} a_{01}) / (\rho_{02} a_{02}) \tag{4.9.7}$$

the index of refraction, we can rewrite the Newton-Snell's law for any two similar, non-standard atmosphere, trajectory:

$$\sin \alpha_{02} = J_{1,2} \sin \alpha_{01}. \tag{4.9.8}$$

The similitude and Newton-Snell's law approach to the solution of exterior ballistics problems is especially very useful in exterior ballistics of small arms to set up the departure angle and the angle of sight, using the individual data obtained in practice of shooting.

Particularly, the use of Newton-Snell's law in exterior ballistics of small arms gives accurate results.

The following example demonstrates the application of Newton-Snell's law and similitude equations to find the range of a projectile based on the range tables, or shooting data observed during firing tests of a given projectile.

Example 9.1 Mountain Shooting

Use the partial range tables of 0.338 GB528 Lapua Bullet to find the departure angle needed to hit the target located at a range of $x_{T2} = 1500 m$ if the firing site is $y = 1500 m$ over the sea level.

- At sea level, the temperature, humidity and pressure are respectively $t = 15°$ Celsius, 0.00% and $p_{0N} = 760 \ mm \ Hg.$, the corresponding virtual temperature is $\tau_{0N} = 288.15°$.

  Launching velocity of the projectile is $v_{01} = 830 \ m/s$.

- At altitude $y = 1500 \ m$ over the sea level, (non-standard atmosphere) the temperature, pressure and, the virtual temperature (humidity 50%) are respectively:

ELEMENTS OF EXTERIOR BALLISTICS | 151

$t = 6°$ Celsius, $p_0 = 626$ mm Hg., $e_{050} = 3.536$ mm Hg., 3
$T_0 = 279.748°K$.

Launching velocity is $v_{02} = 830$ m/s.

The temperature of propellant charge of cartridge, on both shooting sites is the same, 21.11 degree Celsius.

Table 9.1 - 0.338 GB528 Lapua Bullet

Range [m]	700	800	900	1000	1100	1200
Launching Angle	0.3662	0.4367	0.5126	0.5958	0.6858	0.7838
Impact Speed	571	539	507	477	448	421
Time	1.015	1.195	1.386	1.590	1.807	2.037
Drop	- 4.494	- 6.139	- 8.086	- 10.439	- 13.229	- 16.52

Table 9.1 continue

Range [m]	1300	1400	1500
Launching Angle	0.8926	1.013	1.141
Impact Speed	395	371	350
Time	2.283	2.544	2.822
Drop	- 20.368	- 24.858	- 30.074

Solution

Scaling factor is

$$J = \frac{p_{0N}}{p_0}\sqrt{\frac{T_0}{T_{0N}}} = \frac{760}{626}\sqrt{\frac{279.748}{288.15}} = 1.1962.$$

Using (4.9.1), we find the corresponding range of Lapua projectile at 1500 meter standard atmosphere:

$$x_{T1} = J^{-1}x_{T2} = (1.1962)^{-1} \cdot 1500 = 1253.94 \; m.$$

Departure Angle

Using the data of range table of the 0.338 GB528, by interpolation, we find that the departure angle needed to hit the target located at the horizontal range $x_{T1} = 1253.94 m$, is $\alpha_{01} = 0.8425°$.

Substituting in Newton- Snell's law, we find that the departure angle (on the mountain site, 1500 meters over the sea level) is

$$\alpha_{02} = \arcsin(J \cdot \sin \alpha_{01}) = \arcsin[1.1962 \cdot \sin(0.8425)] = 1.008°.$$

Time of Flight

In the same way, using interpolation, we find that the time of flight to $x_{T1} = 1253.94m$ is $t_1 = 2.168s$.

Thus, time of flight to 1500 meters at the shooting site is
$$t_2 = J \cdot t_1 = 1.19622 \cdot (2.168) = 2.594s.$$

Velocity

Using table 9.1, by interpolation, we find that the velocity at $x_{T1} = 1253.94m$ is

$v_{T1} = 406.98m/s$ then
$v_{T2} = v_{T1} = 406.98m/s$

Projectile Drop

Using interpolation, we find that the drop at 1253.94 meter is $\bar{y}_1 = -18.596m$.

So, the drop at 1500 meter is
$$\bar{y}_2 = J^2 \cdot \bar{y}_1 = (1.19622)^2 \cdot (-18.596) = -26.61m.$$

The projectile drop can be estimated using departure angle $\alpha_{02} = 1.008°$:
$$\bar{y}_2 = -1500 \cdot \tan(1.008) = -26.392m$$

Note that numerical integration of differential equations shows that elements of Lapua bullet trajectory, at horizontal range $x_{T2} = 1500m$, at altitude $y = 1500\ m$ are :

Departure angle $\alpha_{02} = 1.007°$, time of flight $t_2 = 2.594s$, velocity $v_{T2} = 406.56m/s$.

Note

Using the similarity equations we use the interpolation technique. It is obvious that the accuracy of prediction will be better if we have range tables for ranges that change every 50 meters or less, i.e. for range 50m, 100m, 150m and so on.

4.10 Tangent Law on Similar Trajectories

Newton-Snell's law (4.9.5) is derived applying the theory of dynamic similarity.

There are two essential facts that can not be explained using only Newton-Snell's law:

1. The Newton-Snell's law is restricted to launching angles α_0 and impact angles α_T, respectively smaller than the critical angles α_{0C} and α_{TC} that correspond to the "total internal reflection" (see example 10.1).
2. The practice of shooting shows that the relation between two corresponding angles of similar trajectories does not depend on the "direction" in which we apply Newton-Snell's law, i.e. from shooting in an atmosphere with greater acoustic impedance to another one with smaller acoustic impedance, or vice versa.
3. Newton-Snell's law (4.9.5) and the equation (4.9.6), respectively
$$\sin \alpha_2 = J \cdot \sin \alpha_1 \text{, and } \cos \alpha_2 = \cos \alpha_1, \quad (4.10.1)$$
are not consistent for relatively large angles.

To make compatible both equations in (4.10.1), we modify the Newton-Snell's law on similar trajectories.

Dividing equations (4.10.1), we can write the "Tangent Law" that relates the corresponding angles of two similar trajectories:
$$\tan \alpha_2 = J \cdot \tan \alpha_1. \quad (4.10.2)$$

All the other similarity equations remain the same
$$x_2 = J \cdot x_1, \quad y_2 = J^2 \cdot y_1. \quad (4.10.3)$$
$$t_2 = J \cdot t_1 \quad v_2 = v_1, \quad (4.10.4)$$

The similarity equations, (4.10.3) - (4.10.4), and the "Tangent Law on Similar trajectories" (4.10.2), allow us to determine accurately the elements of trajectory in non standard-atmosphere using the corresponding elements of the trajectory in standard atmosphere, and vice versa.

Thus, using the range tables that are prepared for a given projectile in standard atmosphere at the sea level (or close to it), we can obtain shooting data or range tables of the same projectile launched in whatever atmosphere.

That is the well-known problem of shooting related with the determination of initial data of shooting in non-standard atmosphere using standard range tables.

Secondly, from the shooting tests performed in non-standard atmosphere, (winter, summer, or high altitude shooting), we can obtain range tables in standard atmosphere.

That is the well-known ballistics problem of "converting firing data to "sea level standard data".

The Tangent Law and the associated set of similitude equations can be used as well to find the departure angle needed to hit a target located at an inclined range.

It is important to note that the similarity equations and "the tangent law on similar trajectories" can be used to find the elements of trajectory no matter whether:

- The projectile is launched in a non-standard atmosphere at sea level, (for example in winter, or summer time), or
- The projectile is fired at a site located over the sea level (mountain shooting, airborne shooting, etc.).

In general, the "tangent law" is valid for any two atmospheres, standard or not. Thus, for two non-standard atmospheres, the first one with density and speed of sound respectively ρ_{01} and a_{01}, and the second one with density and speed of sound respectively ρ_{02} and a_{02}, we can write:

$$\tan \alpha_{02} = J_{1,2} \tan \alpha_{01}, \qquad (4.10.5)$$

where

$$J_{1,2} = (\rho_{01} a_{01}) / (\rho_{02} a_{02}). \qquad (4.10.6)$$

Note that the "tangent law on similar trajectories" (4.10.5), and the related similitude formulae, can be used to construct the range tables of a given projectile in ICAO atmosphere when we know the range tables of the same projectile in ASM atmosphere, or vice versa.

Substituting in (4.10.14):

$$\rho_{01} = 1.2034 kg/m^3, \quad a_{01} = 341.458 m/s \quad \text{(ASM atmosphere)},$$

and

$$\rho_{02} = 1.2251 kg/m^3, \quad a_{02} = 340.30 m/s \quad \text{(ICAO atmosphere)},$$

we find the index of trajectory refraction:

$$J_{1,2} = (1.2034 \cdot 341.458)/(1.2251 \cdot 341.458) = 0.9856.$$

Note The tangent law on similar ballistic trajectories (as well as the Newton-Snell's law) can not be used to construct the range tables of a projectile in ICAO atmosphere when we know the range tables of the same projectile in TSA atmosphere.

Reason

In TSA atmosphere, the temperature of propellant charge is 15 degree Celsius, while the temperature of propellant charge for shooting in ICAO atmosphere (ASM atmosphere) is 21.11 degree Celsius.

The Relevance of the Tangent Law on Similar Projectile Trajectories

The similitude equations (4.10.3) – (4.10.4) express the relationships that exist between the elements of two similar trajectories that have different departure angles but respectively identical velocities at corresponding locations on the trajectories, including the initial velocities and terminal velocities.

In long range shooting, since the angles are relatively small, we can use Newton-Snell's law or Tangent Law.

Thus, for example, using Tangent Law in example 9.1, we find that the departure angle is

$$\alpha_2 = \tan^{-1}(J \cdot \tan \alpha_1) = \tan^{-1}(1.1962 \cdot \tan(0.8425)) = 1.0078°.$$

For relatively small angles, the tangent law can be written:

$$\alpha_2 = J \cdot \alpha_1. \tag{4.10.7}$$

Thus, for the example 9.1, we have

$$\alpha_2 = J \cdot \alpha_1 = 1.1962 \cdot (0.8425°) = 1.0078°$$

Comment

For relatively small angles all similarity equations (4.10.3), (4.10.4), and the Tangent Law (4.10.7) are linearly related.

Compatibility of Similarity Equations and Tangent Law

It is interesting to note the compatibility of the similarity equations

Indeed:

- Dividing the right equation of (4.10.3) with the left one we obtain the tangent law (4.10.2).

- Dividing the first equation of (4.10.3) with the first equation of (4.10.4), for the x-components of velocity we can write

$$v_{x2} = v_{x1} \tag{4.10.9}$$

- Dividing the second equation of (4.10.3) with the first equation of (4.10.4), for the y-components of velocity we can write

$$v_{y2} = J \cdot v_{y1} \tag{4.10.10}$$

- Using (4.10.9) and (4.10.10), for the projectile velocity we have

$$v_2^2 = v_{x2}^2 + v_{y2}^2 = v_{x1}^2 + J^2 \cdot v_{y1}^2. \tag{4.10.11}$$

Thus, we have an equation to estimate the velocity v_2 at a point P_2 on the second similar trajectory, when we know the components of velocity at the corresponding point P_1 on the first trajectory.

The right side of (4.10.11) gives the square of the velocity at the similar point P_1 that corresponds to P_2, i.e.

$$v_1^2 == v_{x1}^2 + J^2 \cdot v_{y1}^2. \tag{4.10.12}$$

Note that the similarity equations and the Tangent Law are valid as far as the departure velocity is standard, i.e. the departure velocity remains standard no matter what is the temperature of air at the shooting site.

To keep the departure velocity standard the cartridge should be stored aat the temperature of standard atmosphere (21.11 degree Celsius in ICAO or ASM atmosphere, and 15 degree Celsius in ASM atmosphere).

Example 10.1 Critical "Reflection" Angle

For similitude trajectories of example 4.1, find the critical angle α_C that corresponds to the "total internal reflection".
Scaling factor is $J = 1.1962$.

Solution

To find the critical angle of "total internal reflection" we have to substitute $\alpha_2 = 90°$ in Newton-Snell's law, $\sin \alpha_2 = J \sin \alpha_1$. Thus we can write:

$$\sin 90° = 1.1962 \cdot \sin \alpha_1.$$

Hence, we find that the critical angle is $\alpha_C = 52.718°$.

We can not apply the Snell's law of ballistic refraction if the departure angle $\alpha_C > 52.7161°$.

Since the impact angle always is bigger than the departure angle, it is obvious that the critical angle is much smaller than the estimated departure critical angle, $\alpha_C = 52.718°$.

Note that in practice, for shooting in non-standard atmosphere, there is no such restriction in departure angle.

Newton-Snell's law can not be applied in problems of exterior ballistics for relatively large angles.

So, to avoid errors, for large departure angles, we must always apply the Tangent Law of trajectory refraction and the related similitude formulae.

5

Differential Equations of Exterior Ballistics

> "The point-mass trajectory approximation is so generally useful, it may well considered to be the backbone of modern exterior ballistics"
>
> Robert L. McCoy[26]

The differential equations of point-mass trajectories can be solved numerically when we know the ballistics coefficient related to G_1-function or G_7- reference functions, or when we know the characteristic G-function of an individual bullet obtained by Doppler radar measurements.

The system of differential equations can be solved using numerical methods. Our PC programs use the Runge-Kutta method of integration of differential equations of variable x and relatively big integration steps, 0.1 meter till around 100 meter.

It is interesting but astonishing to find out that the integration step, we use to solve numerically the differential equations of point-mass projectile trajectory, is relatively big.

The study "Improved Euler's Method Applied in Exterior Ballistics, presented at the end of the chapter, demonstrates and analyses that strange finding.

5.1 Differential Equations of Projectile Trajectory

The standard range tables of small arms are constructed using field shooting tests and solving the system of differential equations that describe mathematically the ballistic trajectory of a point-mass projectile (see my books: EBA, EBNA, EBRM)[27] :

Equations of Variable t, time of flight

$$\begin{cases} \dfrac{dv_x}{dt} = -J^{-1} \cdot c \cdot h(y) \cdot G_D(v) \cdot \dfrac{v_x}{v} \\[6pt] \dfrac{dp}{dt} = -\dfrac{g}{v_x} \\[6pt] \dfrac{dx}{dt} = v_x \\[6pt] \dfrac{dy}{dt} = v_y \end{cases} \qquad (5.1.1)$$

($p = \tan \alpha = dy/dx$)

Equations of Variable x

$$\begin{cases} \dfrac{dv_x}{dx} = -c \cdot J^{-1} \cdot h(y) \cdot \dfrac{G_D(v)}{v} \\[6pt] \dfrac{dp}{dx} = -\dfrac{g}{v_x^2} \\[6pt] \dfrac{dt}{dx} = \dfrac{1}{v_x} \\[6pt] \dfrac{dy}{dx} = p \end{cases} \qquad (5.1.2)$$

where:

$g = 9.80665 m/s^2$ is the constant of gravity,

$$c = \dfrac{i \cdot d^2}{m} 1000 \qquad (5.1.2)$$

is the ballistic coefficient (BC), m and d are respectively the projectile mass and the diameter of projectile, i is the form coefficient of the projectile,

$$G(v) = (3.927 \times 10^{-4} \rho_{0N}) v^2 C_D(v/a_{0N}) \tag{5.1.3}$$

is the function of resistance, $C_D(v/a_{0N})$ is a known reference drag coefficient,

$$h(y) = (\frac{T_0 - 0.006328 y}{T_0})^{4.4} \tag{5.1.4}$$

is the density function that describes the change of relative air density with the height of flying projectile over the shooting ground.

$$J^{-1} = (\frac{p_0}{p_{0N}} \sqrt{\frac{T_{0N}}{T_0}}) \tag{5.1.5}$$

is a scaling factor that depends on the virtual temperature and pressure at the shooting site.

The parameters T_{0N} and p_{0N} are respectively the virtual temperature and pressure of air in standard atmosphere in use, while T_0 and p_0 are the virtual temperature and the pressure at the shooting site.

The BC, in equation (5.1.2), in imperial units (in/lb²) is related to BC in SI unites (m/kg²) by the relation:

$$C = 1.4222/c. \tag{5.1.6}$$

The quantity,

$$\bar{c} = J^{-1} \cdot c, \tag{5.1.7}$$

which appears on the right side of (5.1.1), is the **fictive ballistic coefficient**.

The initial velocity of the projectile is equal to the standard launching velocity of the given firearm and projectile, i.e. v_0.

The scaling factor (5.1.8) can be written:

$$J^{-1} = \frac{p_0 \cdot a_0}{p_{0N} \cdot a_{0N}} \tag{5.1.8}$$

where a_0 and a_{0N} are the sound speeds in respective atmospheres.

Note that we have denoted

$$J^{-1} = 1/J.$$ (5.1.9)

where $J = \rho_{0N} a_{0N} / \rho_0 a_0$.

Note that J^{-1} represents the ratio of **specific acoustic impedances** of air. For the projectile launched at a site with $\rho_0 = \rho_{0N}$ and $a_0 = a_{0N}$ we have $J^{-1} = 1$.

For non-standard atmospheres, $\rho_0 \neq \rho_{0N}$ and $a_0 \neq a_{0N}$, J^{-1} can be less or greater than 1. It becomes zero for the projectile flying in vacuum, $\rho_0 = 0$, $a_0 = 0$.

The specific acoustic impedance is a measure of response (resistance) of the projectile to the applied force.

Thus, J^{-1}, is a characteristic of projectile response to the resistance of air in non-standard atmosphere relative to the standard atmosphere.

Standard Ballistic Trajectory

"The prediction of the standard ballistic trajectory is the primary and fundamental problem of external ballistics"[28].

The term **standard ballistic trajectory** indicates the trajectory of the projectile launched in a standard atmosphere, usually in ICAO atmosphere, where temperature of propellant charge is kept standard, 21.11 degree Celsius (70 degree Fahrenheit).

The range table that corresponds to the standard ballistic trajectory is the **standard range table.**

Solving the system of differential equations (5.1.1) we are able not only to predict the standard ballistic trajectory of a projectile but as well the trajectory of point mass projectile in non-standard atmosphere.

Moreover, introducing the wind in system (5.1.1) and using our own PC programs we can predict the projectile trajectory in presence of ballistic wind.

Ballistic Coefficient and Bullet Mass

Any relative small change in bullet mass (dm/m) changes the BC of bullet according to the equation

$$c_m = c \cdot (1 - dm/m). \tag{5.1.10}$$

Thus, a relative increase of $dm/m = 1\%$ in bullet mass, decreases the BC of GB528 Lapua bullet from $c = 3.796$ into

$$c_m = c \cdot (1 - dm/m) = 3.796 \cdot (1 - .01) = 3.758 m^2 kg.$$

The equation (5.1.10) is considered when we predict the projectile trajectory using PC programs.

5.2 Reference G-Functions of Resistance

Traditionally, in exterior ballistics, to predict the ballistic trajectory, there are used the well known **Reference Standard G-Functions:** G_1, G_2, G_5, G_6, G_7, G_8, G_{43}, G-Siacci etc.

For long range shooting the accuracy of ballistic trajectory that is predicted using a reference standard function and the respective fixed BC is not always accurate, especially for long range shootings.

The form coefficient "i" and ballistics coefficient "c" of a bullet, associated with a reference standard G-function, depend on the type of the reference standard G-function we adopt to predict the ballistic trajectory.

The reference G-functions of resistance that are in use in long range shooting with small arms are mostly reference functions of resistance G_1 and G_7.

The G_1 and G_7 functions depend on the projectile velocity, density of air ρ_{0N} and speed of sound a_{0N}, which are characteristics of standard atmosphere (ICAO, ASM, TSA).

The analytical reference G_1-function and reference G_7-function, derived using (5.1.3), are respectively[29]:

ICAO Atmosphere

$$G_1(v) = \begin{cases} 1.0584 \times 10^{-4} v^2 & v \leq 256 \ m/s \\ 0.315754v - 78.6769 & 256 < v \leq 1000 \ m/s \end{cases}, \quad (5.2.1)$$

and

$$G_7(v) = \begin{cases} 5.7679 \times 10^{-5} v^2 & v \leq 256 \ m/s \\ 0.152593v - 35.1717 & 256 < v \leq 1700 \ m/s \end{cases}. \quad (5.2.2)$$

ASM atmosphere

$$G_1(v) = \begin{cases} 1.00347 \times 10^{-4} \cdot v^2 & \text{for} \quad v \leq 256 m/s \\ 0.312914v - 79.3976 & \text{for} \quad 256 < v \leq 1000 \end{cases}, \quad (5.2.3)$$

and

$$G_7(v) = \begin{cases} 5.66480 \times 10^{-5} \cdot v^2 & \text{for} \quad v \leq 256 m/s \\ 0.150355 \cdot v - 34.7319 & \text{for} \quad 256 m/s < v \leq 1700 \end{cases}. \quad (5.2.4)$$

Ballistic Coefficient

Ballistic coefficient,

$$c = i \cdot \frac{d^2}{m} \cdot 1000, \quad (5.2.5)$$

is related to a given reference G-function and is a characteristic of the projectile (mass m, caliber d, form factor i).

Form factor i and, as result BC of a particular bullet, depend on the reference G-function of resistance, $G_1 = G_1(v)$ or $G_7 = G_7(v)$, as well as on the velocity of projectile.

Thus, if $G(v)$ is the unknown G-function of a particular bullet then the form factor of this bullet with respect to reference G_1-function is

$$i_1 = G(v)/G_1(v), \quad (5.2.6)$$

or

$$i_1 = C(v/a_{0N})/C_1(v/a_{0N}). \quad (5.2.7)$$

The form factor of the given bullet (with unknown drag coefficient) with respect to reference $G_7(v)$ function is

$$i_7 = G(v)/G_7(v), \tag{5.2.8}$$

or

$$i_7 = C(v/a_{0N})/C_7(v/a_{0N}). \tag{5.2.9}$$

As the equations (5.2.6) and (5.2.8) show, the form factor of a particular bullet is not constant but it is a function of the projectile velocity v.

The form factor is nothing but a function of projectile velocity that matches the unknown G-function of a projectile with a given reference function.

Usually, the form factor i is considered constant and equal to an average value, or a value that is obtained experimentally.

5.3 Characteristic G-functions of Resistance

A G-function of resistance $G(v)$ that is unique for a given type of projectile we will call **Characteristic G-function**, while the corresponding drag coefficient $C_D(v/a_{0N})$ we will call **characteristic drag coefficient**.

The above nomination allow us to distinguish a characteristic G-function, $G(v)$, of a projectile from a reference G-function (G_1, G_2, ..., G_7).

Nowadays, the bullet manufacturers are generating characteristic drag coefficients $C_D(v/a_{0N})$ and G-functions of individual bullets

$$G(v) = (3.927 \times 10^{-4} \rho_{0N}) v^2 C_D(v/a_{0N}),$$

using Doppler radar measurements for their bullet.

The trend for long range bullets is the use of characteristic drag coefficients $C_D(v/a_{0N})$ obtained using Doppler radar or other experimental methods.

A series of characteristic drag functions of specific bullets are presented in my book "Exterior Ballistics: The Remarkable Methods", Xlibris, 2014.

In the present book, to predict the elements of the trajectory, we mainly use a characteristic G-functions of air resistance related to an individual bullet.

The form coefficient i, presented in (5.2.5), for a known characteristic G-functions of bullet is one. So, for the corresponding BC we have:

$$c_0 = (1)\frac{d^2}{m}1000. \tag{5.3.1}$$

Another important quality of a characteristic G-function (5.1.3) of a specified bullet is that the Doppler radar (or experimentally) generated drag coefficient $C_D(v/a_{0N})$ accounts as well for the jaw-angle of a spinning projectile.

That means that the characteristic G-function **contains as well the contribution of jaw-angle drag coefficient** $C_{D\delta^2}$.

As result, we expect that the **predicted range and some other calculated elements of the bullet trajectory** (in long range shooting) are as accurate as the corresponding elements of the ballistic trajectory calculated using 6-DOF model or modified point-mass model.

Note that the lateral influence of gyroscopic effect will be introduced as ballistic correction together with the Coriolis lateral and longitudinal effects.

The following table 3.1 is the standard range table of 0.338 Lapua GB528 Scenar 19.44 g. bullet obtained solving the system of equations (5.1.1) employing its characteristic G-function:

$$G(v) = 0.141v - 30.031, \tag{5.3.2}$$

and BC $c_0 = 3.796$.

The standard range table 3.1 is obtained using the PC programs Alapua16.Bas and RLapua16.bas (Appendix D and E).

Table 3.1 – Range Table: Bullet 0.338" Lapua GB528 Scenar 19.44g.

Range [m]	100	200	300	400	500	600
Launching Angle	0.0419	0.0691	0.1353	0.1863	0.242	0.3018
Impact Speed	791	752	714	677	641	606
Time	0.122	0.252	0.388	0.532	0.684	0.844
Drop	-0.074	-0.304	-0.709	-1.307	-2.134	-3.172

Table 3.1 - continue

Range [m]	700	800	900	1000	1100	1200
Launching Angle	0.3662	0.4367	0.5126	0.5958	0.6858	0.7838
Impact Speed	571	539	507	477	448	421
Time	1.015	1.195	1.386	1.590	1.807	2.037
Drop	-4.494	-6.139	-8.086	-10.439	-13.229	-16.52

Table 3. 1 continue

Range [m]	1300	1400	1500
Launching Angle	0.8926	1.013	1.141
Impact Speed	395	371	350
Time	2.283	2.544	2.822
Drop	- 20.368	- 24.858	- 30.074

Hereafter are listed some characteristic G-functions of some bullets.

(1) Characteristic G-function of resistance of Lapua GB528 Scenar 19.44, 8.59 mm, related with ICAO atmosphere is

$$G(v) = \begin{cases} 0.15v - 35.6712 & 280 \leq v \leq 830 \\ 0.02364v - 1.27552 & 136 \leq v < 280 \\ 1.924 \times 10^{-4} v^2 & 0 \leq v < 136 \end{cases}.$$

For the interval of velocities over around 340 m/s we can use the following Characteristic G-function:

$$G(v) = 0.14114 \cdot v - 30.1115, \quad \text{where} \quad 340 < v < 900 \ m/s. \tag{5.3.3}$$

or

$$G(v) = 0.141v - 30.031, \text{ where } 340 \leq v \leq 830 \ m/s. \tag{5.3.4}$$

The mass and caliber of GB528 Scenar bullet are respectively $m = 0.01944$ kg, caliber $d = 0.0086$ m. Departure velocity, $v_0 = 830$ m/s.

The ballistic coefficient is

$$c_0 = \frac{id^2}{m} 1000 = \frac{(1) \cdot (0.00859)^2}{(0.01944)} 1000 = 3.7957 \ m^2/kg. \tag{5.3.5}$$

(2) The Characteristic G-function of resistance of 0.338 Lapua GB488 Scenar 16.2 gr. (mass $m = 0.0162$ kg, diameter $d = 0.0086$ m, ballistic coefficient, $c = 4.5497$, departure velocity, $v = 905$ m/s) in ICAO atmosphere is

$$G(v) = \begin{cases} 7.9381 \times 10^{-5} v^2 & v \leq 256 \\ 0.146787v - 34.4620 & 256 < v \leq 1021 \end{cases}. \tag{5.3.6}$$

For the interval of velocities over around 340 m/s we can use the following G-function:

$$G(v) = 0.138776 \cdot v - 28.676. \tag{5.3.7}$$

where $340\ m/s < v < 1021\ m/s$.

The ballistic coefficient of the bullet is

$$c_0 = \frac{id^2}{m} 1000 = \frac{(1)\cdot(0.00859)^2}{(0.0162)} 1000 = 4.5654 m^2/kg.$$

(3) Characteristic G-function, Caliber 0.308, 168 Grain Sierra International bullet

The G-function of Caliber 0.308, 168 Grain Sierra International bullet (mass, $m = 0.010886\ kg$; diameter, $d = 0.0078232\ m$; BC, $c_0 = 5.6325$; (i=1); muzzle velocity, $v_0 = 792.48\ m/s$.), ICAO atmosphere is

$$G(v) = \begin{cases} 6.73536\times10^{-5} v^2 & v \le 256 \\ 0.179117v - 46.77305 & v > 256 \end{cases}. \quad (5.3.8)$$

(4) Characteristic G-function of Resistance of M118LR Bullet

The characteristic G-function of resistance of M118LR bullet (mass, diameter $d = 0.0078232\ m$, departure velocity, $v_0 = 884\ m/s$, is

$$G(v) = \begin{cases} 4.81097\times10^{-5} v^2 & v \le 256 \\ 0.178659v - 46.77305 & v > 256 \end{cases}$$

Ballistic coefficient is

$$c_0 = \frac{i\cdot d^2}{m} 1000 = \frac{(1)\cdot(0.0078232)^2}{(0.01134)} 1000 = 5.3970\ m^2/kg, \quad (5.3.9)$$

(5) Characteristic G-function of Resistance of 0.50 Caliber, MK211 Bullet

The ICAO atmosphere G-function of resistance of the 0.50 - Caliber MK211 Bullet (mass $m = 0.043441\ kg$, caliber $d = 0.01295\ m$, ballistic coefficient, $c_0 = 3.8627$, departure velocity, $v_0 = 827.53\ m/s$) is

$$G(v) = \begin{cases} 5.14774\times10^{-5} v^2 & v \le 256 \\ 0.185906v - 47.78576 & v > 256 \end{cases}. \quad (5.3.10)$$

(6) Characteristic G-function of Resistance of 300 - Grain .338 - .416 Bullet

The G-function of 300 - Grain .338 - .416 bullet (mass $m = 0.01944$ kg, caliber $d = 0.00859$ m, ballistic coefficient, $c_0 = 3.7914$, departure velocity, $v_0 = 927.40$ m/s), in ICAO atmosphere is

$$G(v) = \begin{cases} 7.07213 \times 10^{-5} v^2 & v \leq 256 \\ 0.149642 v - 34.00946 & v > 256 \end{cases}. \qquad (5.3.11)$$

5.4 Characteristic G-function Versus Reference G-function

The accuracy of projectile trajectory predicted using reference G_1 or G_7 functions of resistance and a fixed ballistic coefficient is not always satisfactory.

To illustrate and explain the discrepancies in projectile trajectory let's consider the vector differential equation that describe the projectile trajectory in standard atmosphere

$$\frac{d\vec{v}}{dt} = \vec{g} - c \cdot h(y) \cdot G_D(v) \frac{\vec{v}}{v}. \qquad (5.4.1)$$

The acceleration of a projectile in flight that results from drag force is

$$\frac{dv}{dt} = -c \cdot h(y) \cdot G_D(v). \qquad (5.4.2)$$

The acceleration (5.4.2) and as result the corresponding trajectory depends on BC and G-function of resistance.

As the above equations show, there are two main reasons why the trajectories predicted using two different G-functions are not the same:

- The characteristic G-function of a given projectile is not identical to the reference G-function (G_1 or G_7).
- A fixed BC of a bullet related to a reference G-function can not adjust the reference G-function to match perfectly the characteristic G-function of the given projectile.

Indeed, let's consider 0.338 Lapua GB528 Scenar 19.44 g. bullet, the reference functions G_1 and G_7, respectively given in formulas (4.2.1) and (4.2.2), as well as the characteristic G-function of resistance given in (4.3.4), i.e.

$$G_1(v) = 0.315754 \cdot v - 78.6769 \qquad (5.4.3)$$

where $256 < v < 1000 m/s$,

$$G_7(v) = 0.152593 \cdot v - 35.1717, \tag{5.4.4}$$

where $246 < v < 1700 m/s$,

$$G(v) = 0.141 \cdot v - 30.031, \tag{5.4.5}$$

where $340 < v < 900 \ m/s$.

In table 4.1 there are shown the values of the G-functions (5.4.5), (5.4.4) and (5.4.3) as function of projectile velocity.

As we can see from the table 4.1, the values of the G_7-function are a close match to the values of characteristic G-function of Lapua bullet.

That is the reason why G_7-function is considered close to the drag function of modern long range bullets.

Form Coefficient and BC

In the last two columns of table 4.1 there are shown the form coefficients i_7 and i_1 estimated using formulas (5.2.8) and (5.2.6), i.e.

$$i_7 = G(v)/G_7(v) \text{ and } i_1 = G(v)/G_1(v).$$

Table 4.1

Range	Velocity	G-Lapua	G_7	G_1	i_7	i_1
0	830	86.999	91.114	185.468	0.951	0.474
100	791	81.500	85.258	172.864	0.953	0.476
200	752	76.001	79.401	160.260	0.955	0.479
300	714	70.643	73.695	147.978	0.957	0.481
400	677	65.426	68.138	136.020	0.960	0.484
500	641	60.350	62.732	124.385	0.963	0.488
600	606	55.415	57.476	113.074	0.967	0.492
700	571	50.480	52.220	101.762	0.972	0.497
800	539	45.968	47.415	91.420	0.976	0.502
900	507	41.456	42.610	81.078	0.983	0.509
1000	477	37.226	38.104	71.382	0.990	0.517
1100	448	33.137	33.750	62.010	0.998	0.528
1200	421	29.330	29.695	53.284	1.009	0.541

1300	395	25.664	25.791	44.881	1.022	0.557
1400	371	22.280	22.186	37.124	1.039	0.579
1500	350	19.319	19.033	30.337	1.059	0.607
					I_7 avrg. 0.9847	I_1 avrg. 0.5133

The average form coefficient of Lapua GB528 bullet with respect to G_7 and G_1 is respectively

$$i_7 = 0.9847, \quad i_1 = 0.5133$$

The corresponding BCs are

$$c_7 = i_7 \cdot \frac{d^2}{m} \cdot 1000 = i_7 \cdot c_0 = (0.9847) \cdot (3.796) = 3.738,$$

and

$$c_1 = i_1 \cdot \frac{d^2}{m} \cdot 1000 = i_1 \cdot c_0 = (0.5133) \cdot (3.796) = 1.948.$$

Note that Wikipedia[30], for GB528 Lapua bullet employs the following BCs: $c_7 = 3.7724$ and $c_1 = 1.8117$.

Lapua Company considers $c_7 = 3.7036 \ m^2/kg$ ($C = 0.384 \ lb/in^2$)[31].

The different values presented above are result of different approximate characteristic G-functions used by different authors (companies).

In table 4.2 are given the drop and the impact velocity of GB528 Lapua bullet predicted by solving system (5.1.1), employing G_7 or G_1, as well as corresponding BCs, $c_7 = 3.738$ and $c_1 = 1.948$.

The predicted values are obtained using the PC program RPROJ16.BAS which employs the reference G-functions (5.4.4) or (5.4.3), (Appendix E).

For comparison, in the last two rows of table 4.2 is given the impact speed and drop presented in External Ballistics[32]

Table 4.2

Range [m]	100	300	600	900	1200	1500
Impact Speed (G1)	787	706	592	493	411	348
Drop (G1)	-0.074	-0.714	-3.222	-8.279	-17.017	-31.017

Impact Speed (G7)	789	710	600	501	416	349
Drop (G7)	-0.074	-0.711	-3.191	-8.157	-16.690	-30.379
Impact Speed	791	711	604	507	422	349
Drop	- 0.074	- 0.715	-3.203	-8.146	- 16.571	-30.035

For long range shooting, table 4.2 shows that the values of velocity and drop, predicted using G_7 and the fixed BC, $c_7 = 3.738$, are satisfactory.

For long ranges, the velocity and drop predicted using G_1 and $c_1 = 1.948$ are not accurate enough.

Median BC Table 4.3

	G7	G1
Mean	0.9847	0.5132
Median	**0.9740**	**0.4995**
St Deviation	0.033	0.0395
Range	0.108	0.132

In table 4.3 there are shown the statistics of form coefficients related with the data of table 4.1.

As we have shown, the projectile trajectory, predicted using G_1 and the fixed BC, $c_1 = 1.948$, is not accurate.

An alternative method to improve the accuracy is to use the median value of the form coefficients instead of the respective mean value, i.e. the form coefficients $i_7 = 0.9740$ and $i_1 = 0.4995$.

The corresponding BCs are respectively

$$c_7 = i_7 \cdot \frac{d^2}{m} 1000 = i_7 \cdot c_0 = (0.9740) \cdot (3.796) = 3.697$$

and

$$c_1 = i_1 \cdot \frac{d^2}{m} 1000 = i_1 \cdot c_0 = (0.4995) \cdot (3.796) = 1.896.$$

In the following table 4.4 are given the Lapua bullet velocity and drop predicted using reference G_7-function, reference G_1-function and the respective median values of BC s, $c_7 = 3.697$, $c_1 = 1.896$.

Table 4.4

Range [m]	100	300	600	900	1200	1500
Impact Speed (G1)	789	709	598	500	419	356
Drop (G1)	- 0.074	- 0.712	- 3.201	-8.187	-16.74	-30.373
Impact Speed (G7)	790	712	602	504	419	352
Drop (G7)	- 0.074	-0.710	-3.183	- 8.121	-16.583	-30.118
Impact Speed	791	711	604	507	422	349
Drop	- 0.074	- 0.715	-3.203	-8.146	- 16.571	-30.035

For comparison, in the last two rows of table 5.4 is given the impact speed and the drop according to end note 2.

The comparison shows that the accuracy in bullet drop and velocity predicted using the median value of BC is satisfactory and very accurate when we use reference G_7-function.

Not only the bullet drops, but also the corresponding velocities are quite equal.

Interval BC

Lapua Company and Sierra use the Interval Ballistics Coefficients to improve the data predicted by using reference G_1-function and a fixed BC[33].

To illustrate the Interval BC, we consider 0.338" GB528 Lapua Scenar bullet and the fixed BC, $c = 1.8117 m^2 / kg$ ($C = 0.785 in^2 / lb.$) that corresponds to G_1 function of resistance [Wikipedia].

To improve the prediction accuracy of the trajectory of 0.338" GB528 Lapua Scenar bullet, we can consider the average form coefficient in smaller intervals of velocities: 830 -714, 714 - 606, 606 - 507, 507 - 421, 421 - 350 (see table 4.1).

In table 4.4 is given the ballistic coefficients versus the velocity interval estimated using the data from table 4.1, i.e. the averages of any four form coefficients.

Thus, for example, using data in table 4.1 we find the form coefficient:

$$i_1 = (0.474 + 0.476 + 0.479 + 0.481) / 4 = 0.4777$$

that corresponds to the interval of velocities 830 - 714.
The interval form coefficients and interval BCs are presented in table 4.5.

In the last two rows of table 4.5 are given the velocity and the bullet drop predicted using the calculated Interval BCs and G_1-function.

Table 4.5 Interval Ballistic coefficient of Lapua GB528 Scenar bullet

Velocity	830 - 714	714 - 606	606 - 507	507 - 421	421 - 350
i_1	0.4777	0.4863	0.500	0.5238	0.5710
BC_1	**1.813**	**1.846**	**1.898**	**1.988**	**2.168**
Range	300	600	900	1200	1500
Impact Speed	714	605	506	420	349
Drop	-0.709	-3.169	-8.07	-16.462	-29.935

It can be easily verified (comparing data of table 4.5 with corresponding data of table 3.1) the satisfactory accuracy of the trajectory predicted using G_1-function and the interval BCs.

Comparing the data of the two last rows of table 4.5 with the corresponding data given in the Wikipedia table, we see that there is a significant improvement due to the interval BC.

Comment

As it is shown above, the variety of fixed BCs (or interval BCs) that can be used to predict with acceptable accuracy the bullet trajectory using G_1 or G_7 reference functions spells out the useless efforts to measure "accurately" a "manufacturer exaggerated BC" [34].

Above all, a BC measured within a short interval of 200 feet, or 300 feet, can not represent the entire ranges of long range shootings.

Estimating Ballistic Coefficient
Method 2

Consider the equation (1.12.5) of EBRM book, i.e. the average BC

$$\overline{c}_k = \frac{\overline{v}_{k-1}^2 \cdot \ln(v_{k-1}/v_k)}{h(y) \cdot G_D(\overline{v}_{k-1}) \cdot (x_k - x_{k-1})} \quad (5.4.6)$$

in the interval (v_{k-1}, v_k). Assume that we know the projectile velocity at different ranges, let's say every 100 meters, from zero till a certain range.

We denote $\overline{v}_{k-1}$ the average velocity in the interval (v_{k-1}, v_k).

For example, for the GB528 Lapua bullet ($G(v) = 0.141 \cdot v - 30.031$) substituting in (5.4.6) we find the average value of BC in the interval of velocities 791-830:

$$c_{100} = \frac{((830+791)/2)^2 \cdot \ln(830/791)}{(1) \cdot (0.141 \cdot (830+791)/2 - 30.031) \cdot (100-0)} = 3.7526$$

Using the velocities shown in table 4.1, we have estimated the ballistic coefficient related to the GB528 Lapua characteristic G-function (third column of table 4.6), G_7-function (column 4) and G_1-function (column 5).

Table 4.6

Range	Velocity	G-Lapua	G_7	G_1
		Ballistic Coefficient		
0.00	830	3.753	3.572	1.784
100	791	3.822	3.645	1.825
200	752	3.800	3.633	1.824
300	714	3.783	3.628	1.826
400	677	3.773	3.629	1.834
500	641	3.771	3.640	1.847
600	606	3.891	3.772	1.923
700	571	3.684	3.588	1.840
800	539	3.830	3.751	1.936
900	507	3.753	3.700	1.926
1000	477	3.814	3.790	1.992
1100	448	3.757	3.770	2.005
1200	421	3.859	3.918	2.116
1300	395	3.836	3.951	2.176
1400	371	3.641	3.817	2.154
1500	350			
	Mean value	3.784	3.682	1.934
	Std. Deviation	0.064	0.079	0.129

In the last two rows of the table 4.6 are shown respectively the average values of BC and the related standard deviation values.

The values of BC we consider to predict the trajectories of the given Lapua bullet are the averages:

$$c_7 = 3.682, \ c_1 = 1.934.$$

ELEMENTS OF EXTERIOR BALLISTICS | 175

5.5 Estimation of Coriolis Effect

The differential equations of point-mass projectile trajectory do not consider the **Coriolis force** exerted on projectile:

$$\vec{F}_C = 2 \cdot m \cdot (\vec{v} \times \vec{\Omega}), \qquad (5.5.1)$$

where $\vec{\Omega}$ is the angular velocity of the Earth that rotates Eastward around its axis (directed North), $\vec{v}$ is the projectile velocity, m projectile mass.

The scalar value of angular velocity of Earth rotation is
$$\Omega = 7.292 \times 10^{-5} \, s^{-1}. \qquad (5.5.2)$$

The Coriolis force modifies the trajectory of projectile curving it in lateral direction.

To the marksman, it appears that the Coriolis force deflects the point-mass projectile to the right of shooting plane when projectile is fired in Northern hemisphere, while in Southern hemisphere the deflection appears to the left.

The Coriolis force changes the range of shooting as well.

The equation (5.5.1) shows that:

- When shooting is in equator (Coriolis force zero), there is not lateral deflection.
- When Shooting is along the meridian the Coriolis lateral deflection is maximum.

Equations of Trajectory in Presence of Coriolis Force

Consider a Cartesian coordinate system where the y-axis is directed along the line that connects the center of the Earth with the muzzle, while x-axis and y-axis are at the horizontal plane at the launching point (muzzle).

The horizontal direction of shooting is along x-axis.

The y-axis makes with the equator plane the angle Λ that is the geodetic latitude.

Consider two other axes: ox_A directed North, and oz_E directed East.

The direction of shooting is determined by the azimuth which is the angle A that ox_A-axis forms with x-axis.

Projections of angular velocity of earth rotation $\vec{\Omega}$ along ox, oy and oz axes are

$$\Omega_x = \Omega \cdot \cos \Lambda \cdot \cos, \quad \Omega_y = \Omega \cdot \sin, \quad \Omega_z = -\Omega \cdot \cos \Lambda \cdot \sin \quad (5.5.3)$$

Considering Coriolis acceleration $\vec{a}_C = 2 \cdot (\vec{v} \times \vec{\Omega})$ and following Shapiro, we obtain the following differential equations that describe the projectile trajectory in presence of drag, gravity and Coriolis force:

$$\begin{cases} \dfrac{dv_x}{dt} = -c \cdot J^{-1} \cdot h(y) \cdot G(v) \cdot \dfrac{v_x}{v} + 2 \cdot \Omega_z \cdot v_y \\[2ex] \dfrac{dv_y}{dt} = -g - c \cdot J^{-1} \cdot h(y) \cdot G(v) \cdot \dfrac{v_y}{v} - 2 \cdot \Omega_z \cdot v_x \\[2ex] \dfrac{dv_z}{dt} = -c \cdot J^{-1} G(v) \cdot \dfrac{v_z}{v} - 2 \cdot (\Omega_x v_y - \Omega_y v_x) \\[2ex] \dfrac{dx}{dt} = v_x, \quad \dfrac{dy}{dt} = v_y, \quad \dfrac{dz}{dt} = v_z, \quad v = \sqrt{v_x^2 + v_y^2} \end{cases} \quad (5.5.4)$$

Initial conditions
$$t = 0, \; x_0 = 0, \; y_0 = 0, \; z_0 = 0,$$

$$v_{x0} = v_0 \cos \alpha_0, \; v_{y0} = v_0 \sin \alpha_0, \; v_{z0} = 0 \quad (5.5.5)$$

where
$g = 9.80665 m/s^2$ is the constant of gravity,

$$c = \frac{i \cdot d^2}{m} 1000 \quad (5.5.6)$$

is the ballistic coefficient (BC), m and d are respectively the projectile mass and the diameter of projectile, the number i is the form coefficient of the projectile,

$$G(v) = (3.927 \times 10^{-4} \rho_{0N}) v^2 C_D(v/a_{0N}) \quad (5.5.7)$$

is the function of resistance,
$$C_D(v/a_{ON}), \tag{5.5.8}$$

is the drag coefficient
$$h(y) = (\frac{\tau_0 - 0.006328y}{\tau_0})^{4.4} \tag{5.5.9}$$

is the relative density function
$$J^{-1} = \frac{p_0}{p_{ON}}\sqrt{\frac{\tau_{ON}}{\tau_0}} \tag{5.5.10}$$

is the scaling factor.

For shooting in a standard atmosphere, ICAO, ASM or TSA, $J^{-1} = 1$.

Special cases

- When shooting is along the meridian, Azimuth $A = 0$, $\sin A = 0$. As result,

$$\Omega_z = -\Omega \cdot \cos \Lambda \cdot \sin A = 0.$$

In this case, the first two equations of system (5.5.4) show that the Coriolis effect does not affect the range of shooting.

The cross range deviation (in z-direction) will be determined by the third equation of (5.5.4).

- If the shooting is in Equator and along it, then latitude $\Lambda = 0$ and Azimuth $A = 90°$. Substituting in (5.5.3) we have

$$\Omega_x = 0, \; \Omega_y = 0, \; \Omega_z = -\Omega.$$

Thus, the second term in the third equation of (5.5.4) is zero.

Since at the time of departure the component $v_z(0) = 0$ and $z = 0$, the last equation of (5.5.4) shows that at any moment the component of velocity along z-axis and the z-coordinate will remain zero at any time. There is no cross deflection of the bullet. The trajectory is on the xoy plane along the equator.

We can solve numerically system (5.5.1) for example, for 0.338 Lapua GB528 Scenar 19.44g:

Ballistic coefficient $c = 3.796$, departure angle $\alpha_0 = 1.1471°$, departure velocity $v_0 = 830 m/s$, G-function of resistance

$$G(v) = 0.141v - 30.031. \tag{5.5.11}$$

5.6 Lift Force and Overturning Moment on Spinning Projectile

A projectile (considered a symmetrical rigid body) in rotational motion around its axis of symmetry, forms an angle δ with the tangent to the trajectory.

The angle δ is called "jaw of repose angle" (fig. 8).

The point mass trajectory model considers $\delta = 0$.

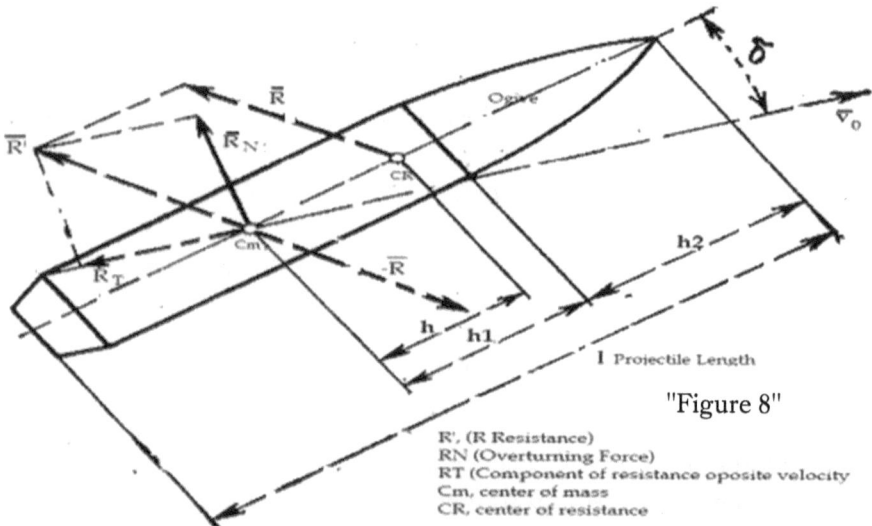

"Figure 8"

R', (R Resistance)
RN (Overturning Force)
RT (Component of resistance oposite velocity
Cm, center of mass
CR, center of resistance

For symmetrical bullets and artillery projectiles, the center of action of drag force $\vec{R}$ is not at the center of mass of projectile C_m, but at a point C_R between the center of mass and the projectile nose. Point C_R is the center of forces of resistance.

The drag force $\vec{R}$ forms an angle Δ with the axis of projectile, which in general is greater than δ, $(\Delta > \delta)$.

As result, the projectile encounters a higher drag force than when we consider $\delta = 0$, i.e., we neglect the lift force.

Since the center of application of the drag force $\vec{R}$ is at the point Cm between center of mass and the nose of projectile, there is a component $\vec{R}_N$ of the drag force $\vec{R}$ that acts in the vertical direction (opposite to the gravity) aiming to overturn the projectile. The vertical force $\vec{R}_N$ is called lift force.

So, there is an overturning moment of the vertical component $\vec{R}_N$ that tends to turn over the projectile.

"Figure 9"

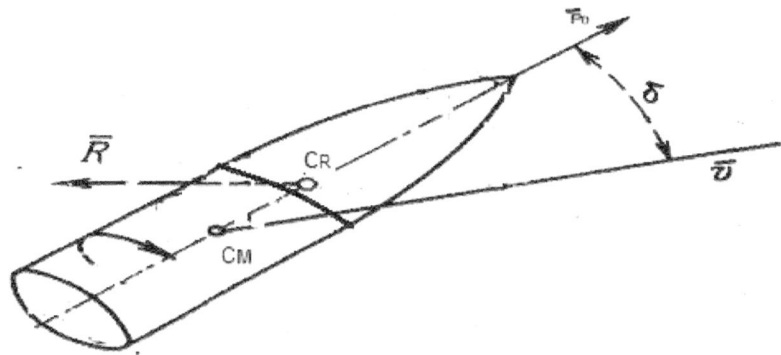

The angular velocity $\vec{p}_0$ of the rotating projectile is along axis of symmetry. Because of the fast rotation of the projectile against its axis of symmetry, the projectile in flights demonstrates gyroscopic properties, i.e. instead of moving vertically, it moves to the direction perpendicular to the plane that contains $\vec{R}_N$ and the angular velocity, i.e. to the right when the rotation of projectile is right-handed, other wise to the left.

The force R_z (in direction of z-axis) exerted to deflect laterally the projectile is equal to the scalar value R_N.

Following Shapiro, and Dmitrievskij for relatively small angles δ, the forces that are applied on the rotating projectile are:

- Drag force in direction opposite to projectile velocity $\vec{v}$, i.e.

$$R = d^2 1000 \cdot h(y) \cdot J^{-1} \cdot \frac{G(v)}{v} \quad (5.6.1)$$

where.
$$G(v) = (3.927 \times 10^{-4} \rho_{0N}) v^2 C_D(v/a_{0N}) \qquad (5.6.2)$$

is the characteristic drag function of projectile, d is reference diameter of projectile, v velocity, ρ_{0N} density of air in standard atmosphere.

- Lift Force (overturning force) directed vertically up, fig. 8.
$$R_N = d \cdot l \cdot h(y) \cdot J^{-1} \cdot \frac{G_L(v)}{v} \cdot \delta, \qquad (5.6.3)$$

where
$$G_L(v) = \frac{\rho_{0N} v^2}{2} C_L\left(\frac{v}{a_{0N}}\right). \qquad (5.6.4)$$

$C_L = C_L(v/a_{0N})$ is the lift (or normal force) force coefficient, l is the projectile length, δ is the angle between $\vec{v}$ and $\vec{p}_0$ (fig. 9).

- The scalar value of moment, $\vec{\mu}$, of overturning force $\vec{R}_N$ (directed perpendicular to the plane that contains velocity $\vec{v}$ and axis of symmetry)
$$\mu = d^2 \cdot h \cdot h(y) \cdot J^{-1} \cdot \frac{G_M(v)}{v} \cdot \delta, \qquad (5.6.5)$$

where
$$G_M(v) = (0.5 \cdot \rho_{0N}) v^2 C_M(v/a_{0N}), \qquad (5.6.6)$$

$C_M(v/a_{0N})$ is the coefficient of overturning moment.

The value of h is estimated using the following formula:
$$h = h_1 + 0.57 h_2 - 0.16 \cdot d, \qquad (5.6.7)$$

h_1 is the distance between the center of mass of projectile and the base of the ogive, while h_2 is the height of the ogive (fig. 8).

For shooting in standard atmosphere $J^{-1} = 1$.

When shooting is in a non standard atmosphere then
$$J^{-1} = \frac{p_0}{p_{0N}} \sqrt{\frac{\tau_{0N}}{\tau_0}}. \qquad (5.6.8)$$

Note that $C_L(v/a_{0N})$ and $C_M(v/a_{0N})$ are functions of Mach number that, for a given projectile, can be determined experimentally.

Example 6.1

Find the value of parameter h for 0.308" 168 grain Sierra International Bullet. Reference diameter $d = 0.308" = 0.007823m$, center of mass from the base $b = 0.474" = 0.01204m$, height of ogive $h_2 = 0.690" = 0.1753m$, projectile length $l = 1.226" = 0.03114m$.

Solution

Referring to fig. 8 we find that
$$h_1 = l - h_2 - b = 0.03114 - 0.01753 - 0.01204 = 0.00157m.$$

Employing equation (5.6.7) we find that
$$h = h_1 + 0.57h_2 - 0.16 \cdot d = 0.00157 + 0.57 \cdot 0.01753 -$$
$$0.16 \cdot 0.007823 = 0.0103m.$$

5.7 Gyroscopic Stability Factor and Twist rate

The characteristics of a symmetrical spinning projectile in flight are the precession and nutation of the axis of symmetry.

As soon as the projectile leaves the muzzle, the spinning projectile performs a conning motion around instantaneous axis of rotation that is along the tangent of trajectory. At the same time the axis of symmetry oscillates (nutation).

The vertex of the cone is at the projectile center of mass.

The angular velocity of precession[35], P, at the moment the projectile is launched, is

$$P = \frac{I_x}{2I_y} p_0, \qquad (5.7.1)$$

where

- I_x is the projectile axial moment of inertia, I_y is the transversal moment of inertia,

$$p_0 = \frac{2\pi \cdot v_0}{\eta \cdot d}, \qquad (5.7.2)$$

is the angular velocity, p_0 (fig.9), of spinning projectile at the moment the projectile is fired, η is the barrel twist of firearm in calibers per turn,

$$\eta = \frac{\pi}{\tan \chi}, \qquad (5.7.3)$$

d reference diameter of bullet or barrel grove diameter, and v_0 initial velocity of projectile, χ is the angle of twist of barrel grooves.

Gyroscopic Stability of projectile

Gyroscopic stability of projectile is related to the spin of the projectile, i.e. with the angular velocity p_0 of rotation (5.7.1) at the moment the projectile leaves the muzzle.

Calculations show that if the angular velocity at the departure point of projectile is over a certain "stability" boundary value, the angular velocity of precession on the remaining part of trajectory will be over that stability boundary value.

When the angular velocity of precession is under the boundary value at the initial moment of shooting, then a slight perturbation will force the projectile to become unstable (tumbling), i.e. the axis of projectile will deviate significantly from the trajectory tangent.

As result, the stability characteristics of trajectory will change extensively and the projectile will miss the target.

Equation (5.7.1) shows that to achieve the gyroscopic stability of projectile we need to increase the angular velocity of spinning ω_0 and reduce the ratio B/A of moments of inertia.

Gyroscopic Stability Criterion

Differential equations[36] that govern the rotation of symmetrical projectile in flight at the initial part of trajectory[37] show that the projectile will be stable if at the instant the projectile leaves the muzzle (departure velocity v_0) the angular velocity (5.7.1), i.e.

$$P_0 = \frac{I_x}{2I_y} p_0, \qquad (5.7.4)$$

where

$$p_0 = \frac{2\pi \cdot v_0}{\eta \cdot d}, \qquad (5.7.5)$$

will satisfy the **gyroscopic stability criterion**:

$$\frac{P_0^2}{M_0} - 1 > 0, \qquad (5.7.6)$$

where

$$M_0 = \frac{d^2 \cdot h}{I_y} \cdot h(y_0) \cdot J^{-1} \cdot G_M(v_0) \qquad (5.7.7)$$

is the angular velocity of precession at the muzzle, while

$$G_M(v_0) = \frac{\rho_{0N} \cdot v_0^2}{2} C_{M0}(\frac{v_0}{a_{0N}}). \qquad (5.7.8)$$

Following Shapiro and Dmitrievskij, the gyroscopic stability factor, denoted S_g, is the number

$$S_g = \frac{P_0^2}{M_0} \qquad (5.7.10)$$

that satisfies the gyroscopic stability criterion:

$$S_g = \frac{P_0^2}{M_0} > 1. \qquad (5.7.11)$$

Hence,

$$S_g = \frac{P_0^2}{M_0} = \frac{2\pi^2 I_x^2}{d^4 h \cdot \rho_{0N} I_y \cdot J^{-1} \eta^2 \cdot C_{Mo}}. \qquad (5.7.12)$$

Note that at the muzzle $h(y_0) = 1$.

If shooting is done in standard atmosphere, $J^{-1} = 1$, then

$$S_g = \frac{P_0^2}{M_0} = \frac{2 \cdot \pi^2 \cdot I_x^2}{d^4 h \cdot \rho_{0N} \cdot I_y \cdot \eta^2 \cdot C_{Mo}}. \qquad (5.7.13)$$

A safety gyroscopic value is considered $S_g = 1.78$, or a value in the interval between 1.5 - 2.00.

From (5.7.1) it follows that the barrel grove twist, η, is related to the gyroscopic stability, S_g, by the equation:

$$\eta = \frac{\sqrt{2} \cdot \pi \cdot I_x}{d^2 \sqrt{h \cdot \rho_{0N} \cdot I_y \cdot J^{-1} S_g \cdot C_{Mo}}}. \qquad (5.7.14)$$

Note

In the American exterior ballistics literature, the stability factor and the barrel twist grove are respectively

$$S_g = \frac{P_0^2}{4M_0} = \frac{\pi^2 \cdot I_x^2}{2 \cdot d^5 \rho_{0N} \cdot I_y \cdot J^{-1} \eta^2 \cdot C_{Mo}} \qquad (5.7.15)$$

and

$$\eta = \frac{\pi \cdot I_x}{d^2 \sqrt{2d \cdot \rho_{0N} \cdot I_y \cdot J^{-1} S_g \cdot C_{Mo}}}. \qquad (5.7.16)$$

Note. The stability criterion of a projectile in flight and most of the formulae in this section are derived in Dynamics of rigid body applying the Lagrange equation to the rotation motion of projectile:

$$T = \frac{1}{2}[I_y(p^2 + q^2) + I_x r^2].$$

We have only shown some results presented in exterior ballistics literature.

For more details the reader should refer to: McCoy's, Modern Exterior Ballistics (chapter 10), Dmitrievskij, Exterior Ballistics, Chapter 10, section 4.

Example 7.1

Find the barrel grove twist for 0.308" for 0.308", 168 grain Sierra International Bullet for shooting in standard atmosphere, $\rho_0 = 1.2251 kg/m^3$.

According to McCoy, reference diameter $d = 0.308" = 0.007823m$, distance of center of mass from the base $b = 0.474" = 0.01204m$, height of ogive $h_2 = 0.690" = 0.1753m$, projectile length $l = 1.226" = 0.03114m$, axial moment of inertia $I_x = 0.000247 lb \cdot in^2 = 7.2282 \times 10^{-8} kg \cdot m^2$, transverse moment of inertia $I_y = 0.001838 lb \cdot in^2 = 5.37872 \times 10^{-7} kg \cdot m^2$. $C_{Mo} = 2.633$ (McCoy, table page 217).

Use $S_g = 1.7$.

Solution

In example (7.1) we found $h = 0.0103m$.

Substituting, we find

$$\eta = \frac{\sqrt{2} \cdot \pi \cdot I_x}{d^2 \sqrt{h \cdot \rho_{0N} \cdot I_y \cdot J^{-1} S_g \cdot C_{Mo}}} =$$

$$= \frac{\sqrt{2} \cdot \pi \cdot 7.2282 \times 10^{-8}}{(0.00782)^2 \sqrt{(0.0103) \cdot (1.2251) \cdot (5.37872 \times 10^{-7}) \cdot (1.7) \cdot (2.633)}} = 30.12$$

The value of barrel grove twist, $\eta = 30.12$, corresponds to the gyroscopic stability factor $S_g = 1.7$.

5.8 Estimation of Projectile Spin Deflection

For a stabilized and rotationally symmetrical projectile, in absence of wind, the differential equations that consider **the gravity, drag and lift force** are[38]

$$\begin{cases} \dfrac{dv_x}{dt} = -c \cdot J^{-1} h(y) \cdot G_D(v) \cdot \dfrac{v_x}{v} \\[6pt] \dfrac{dv_y}{dt} = -g - c \cdot J^{-1} h(y) \cdot G_D(v) \cdot \dfrac{v_y}{v} \\[6pt] \dfrac{dv_z}{dt} = -c \cdot J^{-1} G(v) \cdot \dfrac{v_z}{v} + \dfrac{1}{m} J^{-1} R_z \\[6pt] \dfrac{dx}{dt} = v_x, \quad \dfrac{dy}{dt} = v_y, \quad \dfrac{dz}{dt} = v_z, \quad v = \sqrt{v_x^2 + v_y^2} \end{cases} \quad (5.8.1)$$

- Initial conditions are:
  $t = 0$, $x_0 = 0$, $y_0 = 0$, $z_0 = 0$,
  $v_{x0} = v_0 \cos\alpha_0$, $v_{y0} = v_0 \sin\alpha_0$, $v_{z0} = 0$, (5.8.2)

departure angle α_0, departure velocity $v_0 = 0$.
- v_x, v_y, v_z are the components of velocity $\vec{v}$.
- The gravity acceleration is $g = 9.80665 m/s^2$,
- $c = \dfrac{i \cdot d^2}{m} 1000$ (5.8.3)

is the ballistic coefficient (BC), m and d are respectively the projectile mass and the diameter, i is the form coefficient of the projectile,
- $G_D(v) = (3.927 \times 10^{-4} \rho_{0N}) v^2 C_D(v/a_{0N})$ (5.8.4)

is the function of resistance related to the drag coefficient $C_D(v/a_{0N})$, a_{0N} and ρ_{0N} are respectively the speed of sound and density in standard atmosphere.
- $J^{-1} = (\dfrac{p_0}{p_{0N}} \sqrt{\dfrac{\tau_{0N}}{\tau_0}})$ (5.8.5)

is a scaling factor that depends on the virtual temperature and pressure. τ_o and p_o are the virtual temperature at the shooting site, while τ_{oN} and p_{oN} are respectively the virtual temperature and pressure at standard atmosphere.

- $$h(y) = (\frac{\tau_0 - 0.006328y}{\tau_0})^{4.4} \tag{5.8.6}$$

is the density function.

- $$R_z = \frac{d \cdot l}{m} \cdot h(y) \cdot v^2 \cdot C_L(\frac{v}{a}) \cdot \delta, \tag{5.8.7}$$

is the scalar component of the vector lift force $\vec{R}_L$ directed along oz axis, d is the diameter of projectile, l is the length of projectile, $C_L(v/a)$ is lift drag coefficient while δ is the angle between the projectile velocity $\vec{v}$ and the direction of axis of symmetry of the projectile (angle of nutation).

We can write (5.8.7):

$$R_z = c_L \cdot h(y) \cdot G_L(v) \cdot \delta, \tag{5.8.8}$$

where

$$c_L = \frac{d \cdot l}{m} \tag{5.8.9}$$

is the ballistic coefficient in direction of z-axis,

$$G_N(v) = \frac{\rho_{0N} v^2}{2} C_N(\frac{v}{a_{0N}}) \tag{5.8.10}$$

is the function of resistance of air in direction of z-axis.

As a matter of fact, because of nutation motion, instead of angle δ (in radian) we consider angle δ_p that is the average deviation of the axis of symmetry from the direction of velocity- vector of projectile.

The vector lift force $\vec{R}_L$, acting perpendicular to the trajectory, deviates the centre of mass of the projectile at the direction in which was turned aside the projectile ogive from the shooting plane xoy, i.e. to the right of shooting plane when the spin of projectile rotation is right-handed, otherwise to the left.

Simplified Equations

The system of equations (5.8.1) can be written:

$$\begin{cases} \dfrac{dv_x}{dx} = -c \cdot J^{-1} h(y) \cdot \dfrac{G_D(v)}{v} \\[2ex] \dfrac{dv_y}{dx} = -\dfrac{g}{v_x} - c \cdot J^{-1} h(y) \cdot \dfrac{G_D(v)}{v} \cdot \dfrac{v_y}{v_x} \\[2ex] \dfrac{dv_z}{dx} = -c \cdot J^{-1} \dfrac{G_D(v)}{v} \cdot \dfrac{v_z}{v_x} + c_L J^{-1} \dfrac{G_L(v)}{v_x} \cdot \delta_p \\[2ex] \dfrac{dt}{dx} = \dfrac{1}{v_x}, \; \dfrac{dy}{dx} = \dfrac{v_y}{v_x}, \; \dfrac{dz}{dx} = \dfrac{v_z}{v_x} \end{cases} \quad (5.8.11)$$

If we denote $v_y = p \cdot v_x$ and $v_z = q \cdot v_x$ we can write the above system in the form:

$$\begin{cases} \dfrac{dv_x}{dx} = -c \cdot J^{-1} h(y) \cdot \dfrac{G_D(v)}{v} \\[2ex] \dfrac{dp}{dx} = -\dfrac{g}{v_x^2} \\[2ex] \dfrac{dq}{dx} = c_L J^{-1} \dfrac{G_L(v)}{v_x^2} \cdot \delta_p \\[2ex] \dfrac{dt}{dx} = \dfrac{1}{v_x}, \; \dfrac{dy}{dx} = p, \; \dfrac{dz}{dx} = q \end{cases} \quad (5.8.12)$$

Estimation of angel δ_p

According to Shapiro, the average angle of nutation δ_p (system (5.8.11) and (5.8.12)) can be estimated by the equation:

$$\delta_p = \frac{\pi \cdot g}{2} \cdot \frac{\mu \cdot C_q}{\eta \cdot (h/d)} \cdot v_0 \cdot \frac{\cos \alpha}{h(y) \cdot v^3 \cdot K_M}, \qquad (5.8.13)$$

where:
- η is the twist in caliber, d is the bullet diameter, α angle of flight, v_o is the departure velocity,
- $h = h_1 + 0.57 h_2 - 0.16 \cdot d$, $\qquad (5.8.14)$

h_1 is respectively the distance between the center of mass of projectile and the base of the ogive, while h_2 is the height of the ogive.

- $\mu = \dfrac{4 \cdot I_x}{m \cdot d^2},$ $\qquad (5.8.15)$

I_x is the axial moment of inertia.

- $C_q = \dfrac{m}{1000 \cdot d^3}.$ $\qquad (5.8.16)$

$\delta_p = k \cdot \dfrac{\cos \alpha}{h(y) \cdot v^3 \cdot K_M}$ $\qquad (5.8.17)$

where

$$k = \frac{\pi \cdot g}{2} \cdot \frac{\mu \cdot C_q}{\eta \cdot (h/d)} \cdot v_0 \cdot \qquad (5.8.18)$$

5.9 Improved Euler's Method Applied in Exterior Ballistics

Solution of Differential Equations of Projectile Flight Using Large Steps[39]

George Klimi, PhD

Introduction

In Exterior Ballistics the numerical integration of differential equations of point-mass projectile trajectory, in general, is done using the numerical methods and employing a relatively small step size.

According to Robert McCoy, Euler's method used to solve the point-mass differential equations of projectile trajectory is inaccurate, while the Improved Euler's method is very accurate[40].

McCoy points out as well that at the US Army Ballistics Research Laboratory (Aberdeen Proving Ground, Maryland) the Improved Euler's method (IEM) was the preferred technique to solve the system of differential equations of point-mass projectile.

Note that McCoy, in his MCTRAJ Basic PC Program, uses a step size of 1 yard or 1 meter (McCoy's Modern Exterior Ballistics, page 184).

The C/C++ program IEM.exe[41] (code in Appendix F), and the excel sheet named IEM (appendix F) demonstrate the accuracy of Improved Euler's method (IEM) to solve the Differential Equations of Point-Mass Projectile flying in presence or absence of drag.

Though the step size is unusually big (100 meters) the solution of system of differential equations (5.5.1) is sufficiently accurate. It seems that the truncation errors are insignificant.

An explanation is made employing the Taylor's formula with remainder.

The accuracy of IEM in solving the differential equations of projectile flying in vacuum is perfect even for huge step sizes, 100 - 1500 meters.

The truncation errors as well as the global truncation errors are zero, while the rounding errors are insignificant.

ELEMENTS OF EXTERIOR BALLISTICS | 191

The C++ program IEM.exe, (code in Appendix F) demonstrates the accuracy of the Improved Euler's Method in solving the differential equations[42] of point-mass projectile trajectory (fig. 10),

$$\begin{cases} \dfrac{dv_x}{dx} = -c \cdot h(y) \cdot \dfrac{G_D(v)}{v} \\ \dfrac{dp}{dx} = -\dfrac{g}{v_x^2} \\ \dfrac{dt}{dx} = \dfrac{1}{v_x} \\ \dfrac{dy}{dx} = p \end{cases} \quad (5.9.1)$$

even when the step size is incredibly big, equal to the length of a soccer field.

In system (5.9.1):

$p = \tan \alpha$, where α is the angle the projectile velocity forms with x-axis, (x, y) are the coordinates of projectile at a moment t, v is the projectile velocity (along the tangent to the trajectory, c is the ballistic coefficient (BC), $G_D(v)$ is the function of air resistance of the given projectile and $g = 9.80665 m/s^2$ is the gravity constant.

The component of velocity along x-axis is $v_x = v \cdot \cos \alpha$, while $v_y = v \cdot \sin \alpha$ is the vertical component. Projectile departure angle, at $x = 0$ is α_0.

A series of characteristic $G_D(v)$ functions of air resistance, for some bullets, are given in Appendix B

1. Truncation Errors for the Improved Euler's Method

We consider the trajectory of the **Lapua Scenar GB528 19.44 g (300 gr.)** bullet launched at an angle $\alpha_0 = 1.1471°$ with velocity $v_0 = 830 m/s$ in ICAO atmosphere.

The ballistic coefficient of the given Lapua bullet and the characteristic drag function[43] are respectively $c = 3.796 m^2/kg$ and

$$G_D(v) = 0.141 \cdot v - 30.031. \quad (5.9.2)$$

The density function $h(y)$ at the sea level is 1, i.e. $h(y) = 1$.

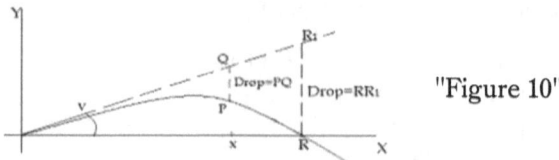

"Figure 10"

The accuracy of the Improved Euler's Method (IEM) is verified comparing the data obtained using the Improved Euler's method (C++ program IEM.exe and IEM excel sheet) and the data in table 9.1, obtained using Doppler radar measurements[44].

Table 9.1. Lapua Scenar GB528 19.44 g (300 gr) bullet

Range (m)	0	300	600	900	1,200	1,500
Velocity (m/s)	830	711	604	507	422	349
Time (s)	0	0.3918	0.8507	1.3937	2.0435	2.8276
Bullet Drop (m)	0	0.715	3.203	8.146	16.571	30.035

In table 9.2, for comparison, are given the corresponding data obtained solving system (5.9.1) by employing the improved Euler's method demonstrated in excel spreadsheet IEM.xls attached (Appendix F).

Table 9.2. Comparison of True Solution with the Approximate Numerical IEM Solution

Range (m)	0		300		600		900		1200		1500	
Velocity (m/s)	830		711	713.79	604	605.23	507	506.49	422	420.27	349	349.45
Time (s)	0		0.3918	0.3900	0.8507	0.847	1.3937	1.3894	2.0435	2.041	2.8276	2.8272
y-coordinate	0		5.292	5.306	8.811	8.861	9.875	9.979	7.457	7.612	0	0.1867
Bullet Drop	0		0.715	0.701	3.203	3.153	8.146	8.042	16.571	16.416	30.035	29.850

From table 9.2, we can see that the error (difference between the true solution and the approximate one) obtained using IEM for the y-coordinate of bullet at 1500 meters is

$$e_{14} = 0.00 - 0.1867 = -0.1867 m.$$

That means that the bullet, at 1500 m, will pass around 19 cm over the center of target. This approximate result is very satisfactory accurate in the practice of long range shooting with small arms.

At range 300 meters the error,
$$e_2 = 5.292 - 5.306 == -0.014m,$$

is insignificant.

Let's estimate the local truncation errors resulting when we use the Improved Euler's method (IEM).

The y-coordinate as a function of range can be expressed using Taylor formula with remainder,

$$y_{n+1} = y_n + y'(x_n)\cdot\frac{h}{1!} + y''(x)\cdot\frac{h^2}{2!} + y^{(3)}(c)\frac{h^3}{3!}, \text{ for } x_n < c < x_{n+1} \quad (5.9.3)$$

where $h = x_{n+1} - x_n$ is the step size.

The sum of the first three terms in (5.9.3) represents the Improved Euler's formula, while the fourth term is the remainder. The remainder is equal to the local truncation error (for the interval $[x_n, x_{n+1}]$) that occurs since we approximate the true solution using limited terms of Taylor series (three terms when we use improved Euler's method).

In other words, we cut off (truncate) a part of the Taylor series solution in $[x_n, x_{n+1}]$ that is estimated using the remainder.

For the projectile trajectory that is described by the system of differential equations (5.9.1) the third derivative[45] of y with respect to x is

$$y^{(3)}(x) = -2g\cdot c\cdot(\frac{G_D(v)/v}{v^3 \cos^3(\alpha)}). \quad (5.9.4)$$

(See formula 2.6.11, p.198, ref. 18).

The remainder in (5.9.3), for the given Lapua bullet (using equation (5.9.2)) is

$$R_n = -2(9.80665)\cdot(3.796)\cdot[\frac{(0.141v_c - 30.031)/v_c}{v_c^3 \cos^3(\alpha_c)}]\frac{h^3}{3!}. \quad (5.9.5)$$

Hence, we can write

$$R_n = -37.22604\cdot[\frac{(0.141v_c - 30.031)/v_c}{v_c^3 \cos^3(\alpha_c)}]\frac{h^3}{3}. \quad (5.9.6)$$

The remainder (5.9.6), in absolute value, is not greater than the error obtained substituting in (5.9.6) instead of v_c and angle α_c respectively the value of velocity and the value of angle at the end, x_{n+1}, of the interval $[x_n, x_{n+1}]$.

For example, using the data obtained in spreadsheet file EulerMethod.xls (velocity $v_c = 349.13615$, cell E143, and $v_c \cos(\alpha_c) = 349.3579$, cell E139) for the interval (1400m, 1500m) we find that the local truncation error is not greater than the absolute value of $R_M = -0.016 m$ (see cell I139 in the excel spreadsheet Appendix F).

Since the global truncation error for IEM theoretically is of the order h^2, i.e. $O(h^2)$, we can say that the accuracy of the numerical integration of system 5.9.1 can be increased by decreasing the step size (though in our case it is not necessary).

Thus, using C++ program for step size 0.1, 1, or 10 we find that the global truncation error at 1500 meters is

$$e_{14} = 0.00 - 0.13 = -0.13 m.$$

Decreasing the step size from 100 to 0.1 we practically have an insignificant improvement of predicted outcome.

The estimation of the truncation errors is valid in general for almost all bullets that are lunched with supersonic velocities till the ranges where the velocity remains supersonic (greater than 340m/s).

For bullets, the IEM assures a great accuracy for relatively large integration steps.

IEM for Point-Mass Projectile in Vacuum

Note that for the ballistics coefficient $c = 0$ the system of differential equations (5.9.1) describes the trajectory of a projectile flying in absence of resistance of air.

The solution of the system of differential equations (5.9.1) with $c = 0$ is the well known parabolic trajectory of the point mass projectile flying in vacuum,

$$y = \tan(\alpha_0) x - \frac{g}{2 v_0^2 \cos^2(\alpha_0)} \cdot x^2, \qquad (5.9.7)$$

where the constant of gravity is $g = 9.0665 m/s^2$.

We can get the Improved Euler's method (IEM) solution of the system (5.9.1) in free space, substituting in cell F12 of the Excel spreadsheet (IEM.xls) $c = 0$ instead of $c = 3.796$.

In this case, we can verify the accuracy of the Improved Euler's method (step 100m) comparing the obtained Excel spreadsheet results with the results that we can find using the equation of parabola (5.9.7).

Thus, for the horizontal range $x = 1500m$, substituting in the equation of parabola (5.9.7):

$g = 9.0665 m/s^2$, departure angle $\alpha_0 = 1.1471°$ and the departure velocity of bullet $v_0 = 830 m/s$, we find the true value $y_T = 14.013966m$.

For the same horizontal range, the spreadsheet file IEM.xls (for $c = 0$) gives the value of $y_a = 14.012648m$ (Cell E142).

The truncation error for the Improved Euler's method ($h = 100m$) is

$e = 14.013966 - 14.012648 = 0.00132m$.

The error above shows that the projectile will fly $0.00132m$ below the center of target located at range 1500m. Practically the bullet will hit the center of the target.

From the practical point of view the numerical integration of system (5.9.1) in absence of air resistance gives a perfect solution for y.

The local truncation error is zero since the third derivative of (5.9.7) is zero.

(Note as well that for $c = 0$ the third derivative of y, given by formula (5.9.4), becomes zero).

So, the remainder of the series is zero and as result, the local truncation errors are zero.

In other words, the numerical integration of the differential equations (5.9.1) of point mass projectile in absence of air resistance is free of local truncation errors.

We can say that **the local truncation error does not depend on the step size**.

It is clear that the **global truncation error is zero** as well.

Indeed, the error (difference between the true value and the approximate value) must be zero. The small error estimated above, $e = 0.00132m$ is a round off error related with the round off errors done in every step.

It is obvious that decreasing the step size, we increase the round off errors since we increase the number of iterations.

In table 9.3 is shown the round off error of the improved Euler's method for the point mass projectile flying in vacuum (step size, 100m). The estimation is done using Excel program.

Table 9.3 Round off Errors (using Excel program)

Range	300	900	1200	1500
Round off Error	$4.6 \times 10^{-5} m$	$2.6 \times 10^{-4} m$	$5.7 \times 10^{-4} m$	$1.3 \times 10^{-3} m$

We see that the round off error increases with range, i.e. with the number of iterations (3 iterations for range 300 meters, 9 for 900m, and 15 for 1500m).

The round off errors, resulting by the use of IEM, for the trajectory of projectile in vacuum (range 1500 meters, step size 100 meters) are practically zero.

Interesting outcome when $c = 0$

Executing the C++ program for c= 0, step size h: 0.1, 1, 10, 100, 300 we get practically the same predicted outcomes for any range till 1500m.

Note: The results obtained using C++ programs (step 100m) are slightly different from the results obtained using Excel program.

Strange outcome for $c = 0$, **step size** $h = 1500$

Executing the C++ program using a huge step size h = 1500 we get practically the same result as above ($e = 14.013966 - 14.0140 = -0.000034m$).

An explanation of the above results is related with the fact that the truncation error, and so the global truncation error, do not depend on the step size.

We can as well affirm that the IEM (Heun's method), as a **predictor-corrector algorithm,** in the case of projectile flying in absence of drag, is in accordance with Physics interpretation that considers the point-mass projectile moving independently and at the same time **along the trajectory tangent** (direction of departure velocity; prediction) and **vertically downwards** (correction), in opposite direction of y-axis, under the action of gravity, i.e. according to the vector equation

$$\vec{r}(t) = \vec{v}_0 t + \frac{\vec{g}}{2} \cdot t^2 \qquad (5.9.8)$$

where $t = x / (v_0 \cos \alpha_0)$, while $\vec{r}(t)$ is the vector-position of the bullet at time t.

Truncation Errors for the Euler's Method

The excel spreadsheet file, in spreadsheet 2, named Excel1.xls (appendix F), as well as the C++ program for step size 100m, demonstrate the solution of system (5.9.1) for the same Lapua bullet using Euler's Method.

For the Euler's method the Taylor formula with remainder is

$$y_{n+1} = y_n + y'(x_n) \cdot \frac{h}{1!} + y''(c) \cdot \frac{h^2}{2!}, \text{ for } x_n < c < x_{n+1} \qquad (5.9.9)$$

where $h = x_{n+1} - x_n$ is the step size.

The sum of first two terms in (5.9.9) represents the Euler's formula, while the third term is the remainder.

For the trajectory of point mass projectile that is described by the system of differential equations (5.9.1), the second derivative of y(x) is

$$y''(x) = -\frac{g}{v^2 \cos^2(\alpha)}. \qquad (5.9.10)$$

(see formula 2.6.10, p.198, ref.18).

For the remainder we can write:

$$R_n = -\frac{g}{v_c^2 \cos^2(\alpha_c)} \cdot \frac{h^2}{2}, \quad x_n < c < x_{n+1} \qquad (5.9.11)$$

The remainder that represents the local truncation error, changes (increases) with the increase of shooting range (because the velocity decreases with range).

We can write:

$$y_{n+1} = y_n + \tan(\alpha_n) h - \frac{g}{2 v_c^2 \cos^2(\alpha_c)} \cdot h^2 \qquad (5.9.12)$$

where α_c and v_c are the projectile velocity and the projectile angle that correspond to the point on the trajectory with abscissa c, ($x_n < c < x_{n+1}$).

The sum of the first two terms in (5.9.12) represents the Euler's formula, while the third term is the local truncation error (5.9.11).

The maximum value of the local truncation error is obtained close to the point on trajectory with abscissa x_{n+1}, where the component of velocity along x-axis, $v_x = v \cos \alpha$, is the smallest on the interval $[x_n, x_{n+1}]$.

For the projectile demonstrated in Excel sheet file, Excel1.xls, the upper boundary of the local truncation error for the last interval (1400m, 1500m) is

$$R_{14} = -\frac{9.80665}{2 \cdot (342.1396)^2} \cdot (100)^2 = -0.4189 m.$$

(For velocity see cell C139, spreadsheet Excel1.xls).

We find that the actual truncation error (see cell C142) for the range 1500 meters is
$$e(h) = y - y^* = 0.00 - 4.1276 = -4.1276 m.$$

That means that the bullet will impact around 4 meters over the center of the target, completely missing it.

The Euler's method solution for the system of equation (5.9.1) is not accurate when we use a large step size of 100 meters.

The upper boundary of the truncation error for the range 300m is

$$R_2 = -\frac{9.80665}{2 \cdot (712.612)^2} \cdot (100)^2 = -0.0966 m.$$

(see cell C43, velocity).

Using the data of the table on the Excel.xls spreadsheet (cell C46 and C179, bullet drop), we find that the error for the range 300 meters is
$$e = y - y^* = 5.292 - 5.5650 = -0.273 m,$$

since the true radar value of y-coordinate of bullet at range 300 meter is
$$y = \tan(1.1471) \cdot 300 - 0.715 = 5.292.$$

Drop equation:
$$\bar{y} = x \cdot \tan(\alpha_0) - PQ$$

For relatively short shooting ranges and strep size 100m, the Euler's method gives acceptable solutions for the practice of long range shooting.

Of course, we can increase the accuracy by decreasing the step size, but probably we risk to increase the round off errors.

Euler's Method for Point-Mass Projectile in Absence of Drag

The system of differential equations (5.9.1), for ballistic coefficient zero, $c = 0$, describes the trajectory of point-mass projectile in absence of air resistance. The exact solution of that system (5.9.1) is given by (5.9.7).

The truncation error for the Euler's method is

$$R_n = -\frac{g}{v_0^2 \cos^2(\alpha_0)} \cdot \frac{h^2}{2}, \quad x_n < c < x_{n+1} \tag{5.9.13}$$

Equation (5.9.13) shows that in each interval, $[x_n, x_{n+1}]$, there is a small but constant local truncation error, i.e.

$$R_n = -\frac{g}{v_0^2 \cos^2(\alpha_0)} \cdot \frac{h^2}{2} = -\frac{9.80665}{830^2 \cdot \cos^2(1.1471)} \cdot \frac{100}{2} = -7.12 \times 10^{-4}.$$

The true error, which is practically mainly result of propagation for x = 1500 meters, is

$$e = y - y^* = 14.013966 - 15.08193 = -1.068 m.$$

The projectile will pass $e = 1.068 m$ over the center of target, completely missing it.

Conclusions

The PC program demonstrations and the analyses of the "big" step of numerical integration is an evidence that the solution of exterior ballistics problems is accurately enough even for huge integration steps.

This contradicts the rule of small steps needed to have accurate results in numerical integration of ordinary differential equations using IEM method.

5.10 Measurement of Muzzle Velocity

The initial velocity v_0, called as well the muzzle velocity, is an important parameter needed to predict the ballistics trajectory of a projectile.

As we have seen in section 1.1, "the **initial velocity** of a bullet (projectile) is a **fictive velocity** that makes the projectile follow the real trajectory if, at the muzzle, the projectile is launched at the **fictive velocity**, assuming that at the muzzle the powder gases cease to increase exit velocity of bullet beyond the muzzle".

The initial velocity of bullets can be measured easily using the chronographs that are already available in the market.

Schematic Design

Two chronographs measure the time Δt_{12} that the bullet passes a short distance Δx_{12} between two screens that are located at least around 4 meter from each other.

The first screen is close to the muzzle in such a distance that the expanding gases can not put it out of order. At the same time, the first screen should be at a distance where the expanding gases have ceased to increase the bullet velocity. That distance, which depends on the firearm and the bullet, can be from few centimeters to some few meters from the muzzle.

For example, Sierra Company describes the measurement of "muzzle" velocity using two screens located around 10 feet from each other. (see http://www.exteriorballistics.com/ebexplained/5th/232.cfm).

The design shows as well a shield screen close to muzzle, to protect the first screen from expanding gases.

That shield might reduce the velocity of bullet, preventing gases to increase the velocity of bullet beyond the shield.

A similar method of measurement is described by Bryan Litz (see Applied Ballistics for Long Range Shooting, 2009, page 33)

As a matter of fact, we measure the average velocity

$$\bar{v} = \frac{\Delta x_{12}}{\Delta t_{12}}, \qquad (5.10.1)$$

ELEMENTS OF EXTERIOR BALLISTICS 201

at the point in the middle of the distance between two screens.

The actual method measures an average velocity at a distance x from the muzzle. Thus, the measured average velocity (5.10.1) is a function of the distance from the muzzle to the midpoint between screens, i.e.

$$\overline{v} = v(x). \tag{5.10.2}$$

The average velocity depends on the midpoint. It is not constant, and so not reliable to represent the muzzle velocity.

Using the measured average velocity (5.10.1) we are able to find the muzzle velocity.

Indeed, let's consider the vector differential equation (2.1.14), that describes the projectile trajectory (see "Exterior Ballistics: The Remarkable Methods", Klimi, G., Xlibris, 2014),

$$\frac{d\vec{v}}{dt} = \vec{g} - c \cdot J^{-1} h(y) \cdot G_D(v) \cdot \frac{\vec{v}}{v}, \tag{5.10.3}$$

where

$$J^{-1} = \frac{p_0}{p_{0N}} \sqrt{\frac{\tau_{0N}}{\tau_0}}, \tag{5.10.4}$$

and

$$G_D(v) = E \cdot v - F. \tag{5.10.5}$$

Some of the G-functions of resistance are shown on Appendix A and Appendix B. For example, for the GB528 Lapua bullet

$$G_D(v) = 0.141 \cdot v - 30.031, \tag{5.10.6}$$

we have $E = 0.141 \cdot v$, and $F = 30.031$.

Since the distance from the muzzle to the mid point between two screens is small, the influence of gravity in projectile trajectory is irrelevant. Thus, the trajectory of the projectile can be considered a straight line, Thus, we can write:

$$\frac{dv}{dt} = -c \cdot J^{-1} h(y) \cdot G_D(v). \tag{5.10.7}$$

For simplicity we consider the ICAO atmosphere. So, $J^{-1} = 1$. The density function is also $h(y) = 1$.

From (5.10.7), we can write

$$\frac{dv}{dx} = -\frac{c \cdot [E \cdot v - F]}{v}. \qquad (5.10.8)$$

Hence

$$-\frac{v \cdot dv}{(E \cdot v - F)} = c \cdot dx. \qquad (5.10.9)$$

Integrating the last equation we find the equation of the trajectory in the following form

$$E \cdot (v_0 - \overline{v}) + \ln(\frac{E \cdot v_0 - F}{E \cdot \overline{v} - F}) = c \cdot E^2 \cdot x, \qquad (5.10.10)$$

where v_0 is the unknown muzzle velocity, x is the distance from the bullet to the midpoint between two screens that can be measured with great accuracy, and $\overline{v}$ is the measured average velocity (5.10.2).

Solving the above equation, for example using a graphing calculator, we find the muzzle velocity.

Formula (5.10.10) can be modified to be used even when the atmosphere is not standard. In a non-standard atmosphere the equation (5.10.10) can be writen

$$E \cdot (v_0 - \overline{v}) + \ln(\frac{E \cdot v_0 - F}{E \cdot \overline{v} - F}) = c \cdot J^{-1} E^2 \cdot x. \qquad (5.10.11)$$

For the non-standard atmosphere the cartridges should be stored at the standard temperature (21.11 degree Celsius for the ICAO and ASM atmosphere).

The equation (5.10.11) can be used to find the muzzle velocity even when the temperature of propellant charge is the same as that of the shooting site.

The muzzle velocity is not the standard velocity of the bullet.

In other words, we can study the dependence of the muzzle velocity on the temperature of cartridge.

NOTE

Equation (5.10.11) can be used also to measure the BC of bullet, let's say relative to the reference G_1 function. In this case, we divide the horizontal range in intervals, let's say each 20 or 50 feet wide, and measure the average velocity in the middle of each interval, using chronometers. Then using (5.10.11) we find the BC in each interval. In this way, we have measured an interval BC.

Example 10.1

Find the muzzle velocity v_0 for the GB528 Lapua Scenar bullet if the average velocity, measured at the midpoint located $x = 6m$ from the muzzle of the firearm is $\overline{v} = 826.83 m/s$.

The BC of the Lapua bullet is $c = 3.796 m^2/kg$. The shooting site is in ICAO atmosphere: Pressure $p_{0N} = 760 mm\ Hg$, humidity 0%, virtual temperature that is equal to the temperature of dry air $\tau_{0N} = 273.15 + 21.11 = 294.26°K$.

Solution

The equation (5.10.10) for the GB528 Lapua Scenar bullet can be written:

$$0.141 \cdot (v_0 - 826.83) + \ln(\frac{0.141 \cdot v_0 - 30.031}{0.141 \cdot (826.83) - 30.031}) =$$

. (5.10.11)

$$= (3.796) \cdot (0.141)^2 \cdot (6)$$

Using a graphing calculator we find that the initial velocity is $v_0 = 829.93 m/s$.

Example 10.2 Non Standard Muzzle velocity

Find the muzzle velocity v_0 for the GB528 Lapua bullet Scenar if at the shooting site the temperature is 12 degree Celsius, pressure 750mm/Hg, humidity 30%.

The average velocity at the midpoint between two screens is $\overline{v} = 819.22 m/s$, the distance from the muzzle to the midpoint is $x = 8.2m$.

The BC of the Lapua bullet is $c = 3.796 m^2/kg$. Use the ICAO atmosphere as reference; Pressure $p_{0N} = 760 mm\ Hg$, humidity 0%, virtual temperature that is equal to the temperature of dry air $\tau_{0N} = 273.15 + 21.11 = 294.26°K$.

Solution

Scaling Factor (see section 3.2)

At shooting site, the temperature in degree Kelvin is

$$T_0 = 273.15 + 12 = 285.15°K.$$

Substituting in (3.2.7), we find that the pressure of air vapors is

$$e_{30\%} = (0.30) \cdot 7.50187 \cdot e^{19.04 \cdot (1. - 280.07/285.15)} = 3.16 mm\ Hg.$$

Using equation (3.2.4), we find the virtual temperature at shooting site

$$\tau_0 = \frac{T_0}{1 - 0.3785 \cdot (e_{30\%}/p_0)} = \frac{285.15}{1 - 0.3785 \cdot (3.16/750)} = 285.61°$$

The scaling factor is

$$J^{-1} = \frac{p_0}{p_{0N}} \sqrt{\frac{\tau_{0N}}{\tau_0}} = \frac{750}{760} \cdot \sqrt{\frac{294.26}{285.61}} = 1.0017.$$

Substituting in (5.10.11), and then solving the obtained equation,

$$0.141 \cdot (v_0 - 819.22) + \ln(\frac{0.141 \cdot v_0 - 30.031}{0.141 \cdot (819.22) - 30.031}) =$$

$$= (3.796) \cdot (1.0017) \cdot (0.141)^2 \cdot (8.2)$$

we find that the initial velocity is $v_0 = 823.48 m/s$.

So, because of the decrease in cartridge temperature, from 21.11 degree Celsius to 12 degree Celsius, the standard muzzle velocity is decreased by

$$\Delta v_0 = 823.48 - 830 = -6.52 m/s.$$

APPENDIX A

Reference G-functions of Resistance

1. BRL G-Functions of Resistance in ICAO Atmosphere

For the ICAO standard atmosphere, the G-function is defined by the equation:

$$G_{DA}(v) = 4.811 \times 10^{-4} \cdot v^2 C_D(v/a_{0N}), \tag{A1}$$

where the speed of sound and the air density at the sea level are respectively:

$$a_{0N} = 340.30 m/s, \quad \rho_{0N} = 1.2251 kg/m^3. \tag{A.2}$$

The density function is

$$h(y) = \left(\frac{288.15 - 0.006328 y}{288.15}\right)^{4.4}. \tag{A.3}$$

Standard Function of Resistance, $G_{1A} = G_{1A}(v)$ - **Type 1 Projectiles**

$$G_{1A}(v) = \begin{cases} 1.0584 \times 10^{-4} \cdot v^2 & \text{for} \quad v \leq 256 m/s \\ 0.315754 \cdot v - 78.6769 & \text{for} \quad 256 < v \leq 1000 \end{cases}, \tag{A.4}$$

or

$$G_{1A}(v) = \begin{cases} 1.0584 \times 10^{-4} \cdot v^2 & for \quad v \leq 256 m/s \\ 0.331547v - 86.0227 & for \quad 256 < v \leq 1200 \end{cases}. \quad (A.5)$$

Standard Function of Resistance, $G_{2A} = G_{2A}(v)$ - Type 2 Projectiles (Usually used in Artillery Fire)

$$G_{2A}(v) = \begin{cases} 9.2868 \times 10^{-5} \cdot v^2 & for \quad v \leq 256 m/s \\ 0.143353 \cdot v - 30.2415 & for \quad 256 < v \leq 1300 m/s \end{cases}, \quad (A.6)$$

or

$$G_{2A}(v) = \begin{cases} 9.2868 \times 10^{-5} \cdot v^2 & for \quad v \leq 256 m/s \\ 0.148718 \cdot v - 33.2882 & for \quad 256 < v \leq 1700 m/s \end{cases}. \quad (A.7)$$

Standard Function of Resistance, $G_{5A} = G_{5A}(v)$ - Type 5 Projectiles

$$G_{5A}(v) = \begin{cases} 7.5244 \times 10^{-5} \cdot v^2 & for \quad v \leq 256 m/s \\ 0.200207 \cdot v - 49.625 & for \quad 256 < v \leq 1500 m/s \end{cases}, \quad (A.8)$$

or

$$G_{5A}(v) = \begin{cases} 7.5244 \times 10^{-5} \cdot v^2 & for \quad v \leq 256 m/s \\ 0.206378 \cdot v - 53.2504 & for \quad 256 < v \leq 1700 m/s \end{cases}. \quad (A.9)$$

Standard Function of Resistance, $G_{6A} = G_{6A}(v)$ - Type 6 Projectiles

$$G_{6A}(v) = \begin{cases} 1.05244 \times 10^{-4} \cdot v^2 & for \quad v \leq 256 m/s \\ 0.142352v - 23.6937 & for \quad 256 < v \leq 1700 m/s \end{cases}. \quad (A.10)$$

Standard Function of Resistance, $G_{7A} = G_{7A}(v)$ - Type 7 Projectiles

$$G_{7A}(v) = \begin{cases} 5.7679 \times 10^{-5} \cdot v^2 & for \quad v \leq 256 m/s \\ 0.152593 \cdot v - 35.1717 & for \quad 256 < v \leq 1700 m/s \end{cases}. \quad (A.11)$$

Standard Function of Resistance, $G_{8A} = G_{8A}(v)$ - Type 8 Projectiles (Employed in artillery fire, and small arms)

$$G_{8A}(v) = \begin{cases} 1.01154 \times 10^{-4} \cdot v^2 & for \quad v \leq 256 m/s \\ 0.149441 \cdot v - 29.3790 & for \quad 256 < v \leq 1500 m/s \end{cases}, \quad (A.12)$$

or

$$G_{8A}(v) = \begin{cases} 1.01154 \times 10^{-4} \cdot v^2 & \text{for} \quad v \leq 256 m/s \\ 0.152307 \cdot v - 31.0336 & \text{for} \quad 256 < v \leq 1700 m/s \end{cases}. \quad \text{(A.13)}$$

The Spherical Projectile G-Function of Resistance, $G_{SA} = G_{SA}(v)$

$$G_{SA}(v) = 2.7189 \times 10^{-4} v^2, \text{ for } v \leq 1400 m/s. \quad \text{(A.14)}$$

According to McCoy, the roughness of surface of the sphere "causes slight drag increase at all speeds", while the very rough spheres causes a non-negligible increase of 10% of the drag at subsonic speeds and around 5% at supersonic speeds. Thus, the (A.14) can be used as well for spherical projectiles of moderate roughness.

The Cylindrical Projectile G-Function of Resistance, $G_{CA} = G_{CA}(v)$

$$G_{CA}(u) = 2.55078 \times 10^{-4} v^2, \text{ for } 150 < v \leq 1200 m/s. \quad \text{(A.15)}$$

2. BRL G-Functions of Resistance in ASM Atmosphere

In ASM standard atmosphere (the air density at the sea level, $\rho_{0N} = 1.2034 kg/m^3$), the G-function is

$$G_D(v) = (4.726 \times 10^{-4}) v^2 C_D(\frac{v}{a_{0N}}), \quad \text{(A.16)}$$

where $a_{0N} = 341.458 m/s$ is the speed of sound at the sea level.

The density function in ASM standard atmosphere is

$$h(y) = (\frac{289.6 - 0.006328 y}{289.6})^{4.4}. \quad \text{(A.17)}$$

Standard Function of Resistance, $G_1 = G_1(v)$ **- Type 1 Projectiles**

$$G_1(v) = \begin{cases} 1.00347 \times 10^{-4} \cdot v^2 & \text{for} \quad v \leq 256 m/s \\ 0.312914 v - 79.3976 & \text{for} \quad 256 < v \leq 1000 \end{cases}, \quad \text{(A.18)}$$

or

$$G_1(v) = \begin{cases} 1.00347 \times 10^{-4} \cdot v^2 & \text{for} \quad v \leq 256 m/s \\ 0.332044 v - 87.6845 & \text{for} \quad 256 < v \leq 1200 \end{cases}. \quad \text{(A.19)}$$

Standard Function of Resistance, $G_2 = G_2(v)$ - Type 2 Projectiles

$$G_2(v) = \begin{cases} 9.116233 \times 10^{-5} \cdot v^2 & \text{for} \quad v \leq 256 m/s \\ 0.141300 \cdot v - 29.9097 & \text{for} \quad 256 < v \leq 1300 m/s \end{cases}, \quad (A.20)$$

or

$$G_2(v) = \begin{cases} 9.116233 \times 10^{-5} \cdot v^2 & \text{for} \quad v \leq 256 m/s \\ 0.146467 \cdot v - 32.7991 & \text{for} \quad 256 < v \leq 1700 m/s \end{cases}. \quad (A.21)$$

Standard Function of Resistance, $G_5 = G_5(v)$ - Type 5 Projectiles

$$G_5(v) = \begin{cases} 7.43571 \times 10^{-5} \cdot v^2 & \text{for} \quad v \leq 256 m/s \\ 0.19734 \cdot v - 49.0806 & \text{for} \quad 256 < v \leq 1500 m/s \end{cases}, \quad (A.22)$$

$$G_5(v) = \begin{cases} 7.43571 \times 10^{-5} \cdot v^2 & \text{for} \quad v \leq 256 m/s \\ 0.203422 \cdot v - 52.6662 & \text{for} \quad 256 < v \leq 1700 m/s \end{cases}. \quad (A.23)$$

Standard Function of Resistance, $G_6 = G_6(v)$ - Type 6 Projectiles

$$G_6(v) = \begin{cases} 1.0334 \times 10^{-4} \cdot v^2 & \text{for} \quad v \leq 256 m/s \\ 0.140533 \cdot v - 23.6633 & \text{for} \quad 256 < v \leq 1700 m/s \end{cases}. \quad (A.24)$$

Standard Function of Resistance, $G_7 = G_7(v)$ - Type 7 Projectiles

$$G_7(v) = \begin{cases} 5.66480 \times 10^{-5} \cdot v^2 & \text{for} \quad v \leq 256 m/s \\ 0.150355 \cdot v - 34.7319 & \text{for} \quad 256 m/s < v \leq 1700 \end{cases}. \quad (A.25)$$

Standard Function of Resistance, $G_8 = G_8(v)$ - Type 8 Projectiles

$$G_8(v) = \begin{cases} 9.9366 \times 10^{-5} \cdot v^2 & \text{for} \quad v \leq 256 m/s \\ 0.14733 \cdot v - 29.0850 & \text{for} \quad 256 < v \leq 1500 m/s \end{cases}, \quad (A.26)$$

or

$$G_8(v) = \begin{cases} 9.9366 \times 10^{-5} \cdot v^2 & \text{for} \quad v \leq 256 m/s \\ 0.150101 \cdot v - 30.6672 & \text{for} \quad 256 < v \leq 1700 m/s \end{cases}. \quad (A.27)$$

The Spherical Projectile G-Function of Resistance, $v \leq 1400 m/s$.

$$G_S(v) = 2.60444956 \times 10^{-4} v^2. \quad (A.28)$$

ELEMENTS OF EXTERIOR BALLISTICS

The Cylindrical Projectile G-Function of Resistance, $150 < v \leq 1200 m/s$

$$G_C(v) = 2.5117 \times 10^{-4} v^2. \tag{A.29}$$

3. G-Function of Resistance in TSA, ICAO, and ASM Atmosphere

In TSA standard atmosphere (the air density at the sea level, $\rho_{0N} = 1.2034 kg/m^3$), the G-function is

$$G_D(v) = 4.7320 \times 10^{-4} \cdot v^2 \cdot C_D(v/a_{0N}), \tag{A.30}$$

where $a_{0N} = 340.84\ m/s$ is the speed of sound at the sea level.

The density function in ASM standard atmosphere is

$$h(y) = (\frac{289.08 - 0.006328 y}{289.08})^{4.4}. \tag{A.31}$$

The Russian G-function of Year 1943 (TSA atmosphere)

The G-function in use by the Russian army is the so-called law of resistance of 1943:

$$G_{43}(v) = \begin{cases} 7.4542 \times 10^{-5} v^2 & v \leq 271\ m/s \\ 0.187337 v - 50.3858 & 271 < v \leq 1400 \\ 1.2313 \times 10^{-4} v^2 & 1400 < v \leq 1700 \end{cases}, \tag{A.32}$$

or

$$G_{43c}(v) = \begin{cases} 7.4542 \times 10^{-5} v^2 & v \leq 256 \\ 0.1157713 v - 36.39542 & 256 < v \leq 1400 \\ 1.2315 \times 10^{-4} & v > 1400 \end{cases}. \tag{A.33}$$

The latest is used in EBNA (Xlibris, 2010).

Approximate G_{43} - function (TSA)

$$G_{43a}(v) = \begin{cases} 7.4542 \times 10^{-5} v^2 & v \leq 256 \\ 0.158738 v - 36.8789 & 256 < v \leq 900 \end{cases}. \tag{A.34}$$

G_{43} - function in ICAO atmosphere

$$G_{43I}(v) = \begin{cases} 7.535 \times 10^{-5} v^2 & v < 271 \\ 0.165114v - 39.2361 & 271 \leq v \leq 1020 \end{cases}. \quad (A.35)$$

G_{43} -function in ASM atmosphere is

$$G_{43A}(v) = \begin{cases} 7.4482 \times 10^{-5} v^2 & v < 271 \\ 0.163211v - 38.7839 & 271 \leq v \leq 1020 \end{cases}. \quad (A.36)$$

Siacci's Original Analytical Function (TSA atmosphere)

$$K_S(v) = 0.2002 \cdot v - 48.05 + [(0.1648v - 47.95)^2 + 9.6]^{1/2} +$$

$$\frac{0.0442 \cdot v(v - 300)}{371 + (v/200)^{10}}. \quad (A.37)$$

Approximate Siacci's G-function (TSA atmosphere)

$$G_{ST}(v) = \begin{cases} 1.2094 \times 10^{-4} v^2 & v \leq 257.4 \\ 0.362013v - 93.1823 & 257.4 < v \leq 1200 \end{cases}. \quad (A.38)$$

Approximate Siacci's, G_{SI} - function of resistance, in ICAO atmosphere

$$G_{SI}(v) = \begin{cases} 1.2296 \times 10^{-4} v^2 & v \leq 257.4 \\ 0.36874v - 95.2603 & 257.4 < v \leq 1200 \end{cases}. \quad (A.39)$$

Approximate Siacci's G_{SA} - function of resistance, in ASM atmosphere

$$G_{SA}(v) = \begin{cases} 1.2078 \times 10^{-4} v^2 & v \leq 257.4 \\ 0.36223v - 93.5778 & 257.4 < v \leq 1200 \end{cases}. \quad (A.40)$$

Approximate Siacci's G-function, (TSA, used in EBA, Xlibris 2008)

$$K_D(v) = \begin{cases} 1.212 \cdot 10^{-4} v^2 & \text{for} \quad v \leq 256 m/s \\ (v - 240)/3 & \text{for} \quad 256 < v \leq 900 \ m/s \end{cases}. \quad (A.41)$$

APPENDIX B

Characteristic G-functions

1. Characteristic G-function of resistance of 0.338 GB528 Lapua Scenar 19.44, 8.59 mm, velocity 830 m/s, BC = 3.796

(ICAO atmosphere):
$$G_D(v) = \begin{cases} 0.141 \cdot v - 30.031 & 325 \leq v \leq 850 \\ 0.02438 \cdot v - 1.3696 & v < 325 \end{cases} \quad (B1)$$

(ASM atmosphere):
$$G_D(v) = \begin{cases} 0.13895 \cdot v - 29.709 & 325 \leq v \leq 850 \\ 0.0242 \cdot v - 1.399 & v < 325 \end{cases} \quad (B2)$$

2. Characteristic G-function of resistance of 0.300 Winchester Magnum Bullet, ($m = 0.012312 \ kg$, caliber $d = 0.0078232 \ m$, departure velocity, $v_0 = 884 \ m/s$, form coefficient $i = 1$, ballistic coefficient $c = 4.97096 \ m^2/kg$,

(ICAO atmosphere):

$$G_D(v) = \begin{cases} 6.49481 \times 10^{-5} v^2 & v < 270 \\ 0.182304v - 44.271043 & v \geq 270 \end{cases} \quad (B3)$$

ASM Atmosphere

$$G_D(v) = \begin{cases} 6.3801 \times 10^{-5} v^2 & v \leq 256 \\ 0.17944v - 43.592056 & v > 256 \end{cases} \quad (B4)$$

TSA Atmosphere

$$G_D(v) = \begin{cases} 6.40817 \times 10^{-5} v^2 & v \leq 256 \\ 0.179871v - 43.68047 & v > 256 \end{cases} \quad (B5)$$

3. Characteristic G-function of resistance of M118 LR bullet (Long Range, sniper bullet).: mass $m = 0.01134 kg$, caliber $d = 0.0078232 \ m$, departure velocity, $v_0 = 884 \ m/s$, ballistic coefficient, $BC = 5.3970 \ m^2/kg$.

ICAO atmosphere:

$$G_D(v) = \begin{cases} 4.81097 \times 10^{-5} v^2 & v \leq 256 \\ 0.178659v - 46.77305 & v > 256 \end{cases}. \quad (B6)$$

4. G-function of Resistance of M118 Ball bullet

The characteristic G-function of resistance of M118 Ball Bullet (Federal GM308M2. (mass $m = 0.01134 \ kg$, caliber $d = 0.0078232 \ m$, ballistic coefficient $c = 5.3973$, departure velocity, $v_0 = 792.48 \ m/s$).

ICAO atmosphere

$$G_D(v) = \begin{cases} 6.39859 \times 10^{-5} v^2 & v \leq 256 \\ 0.181256v - 44.59324 & v > 256 \end{cases}. \quad (B7)$$

Note The BC of the given bullet related with G_7-function is $c = 5.9258 \ m^2/kg$.

ELEMENTS OF EXTERIOR BALLISTICS | 213

5. Characteristic G-function of Resistance of 300 - Grain .338 - .416 Bullet

(mass $m = 0.01944$ kg, caliber $d = 0.00859$ m, ballistic coefficient, $c = 3.7914$, departure velocity, $v_0 = 927.40$ m/s)

ICAO atmosphere

$$G_D(v) = \begin{cases} 7.07213 \times 10^{-5} v^2 & v \leq 256 \\ 0.149642v - 34.00946 & v > 256 \end{cases}. \tag{B8}$$

6. G-function of Resistance of Caliber 0.30 Ball M_2 bullet

The characteristic G-function of Caliber 0.30 Ball M_2, ((mass, $m = 0.00972$ kg; diameter, $d = 0.0078232$ m; $c = 6.2965$; departure velocity, $v = 853.440$ m/s),

ICAO atmosphere is

$$G_D(v) = \begin{cases} 1.2027 \times 10^{-4} v^2 & v \leq 256 \\ 0.168240v - 35.7491 & v > 256 \end{cases}, \tag{B9}$$

Average ballistic coefficient with respect to reference G_7 function:

$c_7 = 3.7149$ with respect to G_7-function ($i_7 = 0.590$).

7. Characteristic G-function of Caliber 0.308, 168 Grain Sierra International bullet:

(mass, $m = 0.010886$ kg; diameter, $d = 0.0078232$ m; $c = 5.6325$; (i=1); muzzle velocity, $v_0 = 792.48$ m/s.).

ICAO atmosphere

$$G_D(v) = \begin{cases} 6.73536 \times 10^{-5} v^2 & v \leq 256 \\ 0.179117v - 46.77305 & v > 256 \end{cases}. \tag{B10}$$

NOTE. All the PC programs can be modified to predict the trajectories of projectiles for which are known the characteristic G-functions

APPENDIX C

QBasic PC Program BC2016.BAS
Finding Ballistics Coefficient Automatically
Ballistic Coefficient (BC): c=1000id^2/m
Non - Standard Atmosphere or Standard
The computed coefficient "ko" is reserved in the file c:/koef.dat

```
'----------------------------------------------------------------------
'Control Example:
'Estimate the BC of 0.338 G528 Lapua Scenar bullet corresponding to G7 function:
'Initial speed = 830m/s; departure angle = 1.1409 degree; Range = 1500m
'Atmosphere: ICAO

'SOLUTION PROCEDURE
'Input Data
'Cancel data File : Print y/n
'Input "Atmosphere: ASM = 1, ICAO = 2; Standard = 3"; 2
'Input "G-Function; G1= 1, G2=2, G5=5, G6=6, G7=7, G8=8"; 7

'INPUT "Guess Initial BC = "; 1
'INPUT "Departure Angle [Degree] "; 1.141
'INPUT " Initial sped [m/s] "; 830
'INPUT "x-coordinate of Target [m] = "; 1500
'INPUT "Y-coordinate of Target [m] "; 0
'INPUT "y-coordinate of Muzzle "; 0
'INPUT "Number of Steps n "; 10
```

ELEMENTS OF EXTERIOR BALLISTICS | 215

'INPUT "Error in x-coordinate "; 0.2
'INPUT "Temperature at Firing Site "; 15
'INPUT " Temperature of Propellant "; 21.11
'INPUT "Pressure at Firing Site "; 760
'Input "Humidity (Decimal) "; 0

'Output:
'Ballistics Coefficient BC = 3.6842
'X-coordinate of Target/Range = 1500
'Departure angle = 1.141
'Impact Speed = 352.97
'Time = 2.8258
'Impact angle =-2.032 degree
'Error in y-coordinate = 0.007
'Error in x-coordinate = 0.2

'0.338 LAPUA GB528 BC with Respect to G1 Reference Standard Function, ICAO Atmosphere

'INPUT "Guess Initial BC = "; 1
'INPUT "Departure Angle [Degree] "; 1.141
'INPUT " Initial speed [m/s] "; 830
'INPUT "x-coordinate of Target [m] = "; 1500
'INPUT "Y-coordinate of Target [m] "; 0
'INPUT "y-coordinate of GUN "; 0
'INPUT "Number of Steps n "; 10
'INPUT "Error in x-coordinate "; 0.2
'INPUT "Temperature at Firing Site "; 15
'INPUT " Temperature of Propellant "; 21.11
'INPUT "Pressure at Firing Site "; 760
'Input "Humidity (Decimal) "; 0.00

'Output:
'Ballistics Coefficient BC = 1.8683
'X-coordinate of Target/Range = 1500
'Departure angle = 1.1471
'Impact Speed = 359.42
'Time = 2.88178
'Angle = - 2.0059
'Error in y-coordinate = 0.007
'Error in x-coordinate = 0.2

'--

'Functions, Subs

DECLARE SUB y1z1v1w1 (x, y, z, v, w, y1, z1, v1, w1, koef, E, F, D, TE, pa1, bb, ta1, Pr, pa)

```
DECLARE SUB InfHyres (koef, kk, dis, n, speed, gab, yy, vo, G, GA, GS, E, F, D, atm,
TE, ta, tp, pa, ea, pa1, Pr, Tc)
DECLARE SUB NPxyzvw (nk, x, x0, y, y0, z, z0, v, v0, w, w0, h, h0, k, l, r, Q)
DECLARE SUB NPkoef (k, l, r, Q, h, y1, z1, v1, w1)
DECLARE SUB menu (cog, cof, xf, yf, xfu, yfu, t$)
DECLARE SUB Rezervim (koef, kek, gab, x0)
DECLARE SUB KthimiKendit (kk)

'Variables

DIM m(4, 4), v(4)
rendi = 4
cog = 7: cof = 0
gab = gab
dkoef = .0001
menu cog, cof, 3, 10, 21, 70, "INPUT"
InfHyres koef, kk, dis, n, speed, gab, yy, vo, G, GA, GS, E, F, D, atm, TE, ta, tp, pa,
ea, pa1, Pr, Tc
CLS
PRINT "First Angle="; kk
hap = n
KthimiKendit kk

'Solution
ff:
x0 = 0: y0 = kk: z0 = y0: v0 = vo: w0 = 0: xx = dis: h0 = hap
y0 = speed * COS(y0 * 3.141516954# / 180)
z0 = TAN(z0 * 3.141516954# / 180)
F:
FOR nk = 1 TO rendi
   NPxyzvw nk, x, x0, y, y0, z, z0, v, v0, w, w0, h, h0, k, l, r, Q
   y1z1v1w1 x, y, z, v, w, y1, z1, v1, w1, koef, E, F, D, TE, pa1, bb, ta1, Pr, pa
   NPkoef k, l, r, Q, h, y1, z1, v1, w1
   m(nk, 1) = k: m(nk, 2) = l
   m(nk, 3) = r: m(nk, 4) = Q
NEXT nk


FOR i = 1 TO rendi
   v(i) = 1 / 6 * (m(1, i) + 2 * m(2, i) + 2 * m(3, i) + m(4, i))
NEXT i

'New Point

x0 = x0 + h: y0 = y0 + v(1): z0 = z0 + v(2)
v0 = v0 + v(3): w0 = w0 + v(4)
```

```
IF ABS(x0 - xx) <= .01 AND v0 <= (yy + gab * TAN(kk * 3.1415 / 180)) AND v0 >=
(yy + (-1 * gab) * TAN(kk * 3.1415 / 180)) AND ABS(x0 - xx) <= .01 THEN
  'Display results
  kek = gab * z0
  impact = 180 * ATN(z0) / 3.141592654#
  CLS
  LOCATE 7, 21: PRINT «Ballistics Coefficient BC = «; koef
  LOCATE 8, 21: PRINT "Range = "; x0
  LOCATE 9, 21: PRINT "Departure Angle = "; kk
  LOCATE 10, 21: PRINT "Impact Speeed = "; y0
  LOCATE 11, 21: PRINT "Time of Flight = "; w0
  LOCATE 12, 21: PRINT "Impact Angle = "; impact
  LOCATE 13, 21: PRINT "Error in y-coordinate = "; ABS(kek)
  LOCATE 14, 21: PRINT "Error in x-coordinate = "; ABS(gab)

  PLAY "a8a16b8a8"
  Rezervim koef, kek, gab, x0
  dis = dis + 200
  IF dis > disf THEN PRINT "END": INPUT b: GOTO fundi:
  hap = n
  READ kk
  PRINT "New Angle ="; kk
  KthimiKendit kk
  GOTO ff:
  fundi:
  END
END IF

dkoef = .0001

IF ABS(x0 - xx) <= .01 AND v0 > (gab * TAN(kk * 3.1415 / 180)) THEN

  koef = koef + dkoef
  PRINT v0, koef
  GOTO ff:
END IF
IF ABS(x0 - xx) <= .01 AND v0 < ((-1 * gab) * TAN(kk * 3.1415 / 180)) THEN

  koef = koef - dkoef
  PRINT v0, koef
  GOTO ff:
END IF
GOTO F:

END
```

```
SUB InfHyres (koef, kk, dis, n, speed, gab, yy, vo, G, GA, GS, E, F, D, atm, TE, ta, tp,
pa, ea, pa1, Pr, Tc)
CLS

LOCATE 4, 12: INPUT "Atmosphere: ASM = 1, ICAO = 2; Traditional Standard
TSA = 3"; atm
CLS
IF atm = 1 THEN GOTO 100:
IF atm = 2 THEN GOTO 200:
IF atm = 3 THEN GOTO 300:

100 LOCATE 4, 13: INPUT "G-Function; G1=1, G2=2, G5=5, G6=6, G7=7, G8=8"; G
TE = 289.6: Pr = 750: Tc = 21.11
IF G = 1 THEN E = .312914: F = 79.3976: D = .000100347#
IF G = 2 THEN E = .1413: F = 29.9097: D = .00009116233#
IF G = 5 THEN E = .19734: F = 49.0806: D = .0000743571#
IF G = 6 THEN E = .140533: F = 23.6633: D = .00010334#
IF G = 7 THEN E = .150355: F = 34.7319: D = .000056648#
IF G = 8 THEN E = .14733: F = 29.085: D = .000099366#

CLS
GOTO 400:

200 LOCATE 4, 13: INPUT "G-Function; G1=1, G2=2, G5=5, G6=6, G7=7, G8=8"; GA
TE = 288.15: Pr = 760: Tc = 21.11
IF GA = 1 THEN E = .31574: F = 78.6769: D = .00010584#
IF GA = 2 THEN E = .143353: F = 30.2415: D = .00009287#
IF GA = 5 THEN E = .200207: F = 49.625: D = .000075244#
IF GA = 6 THEN E = .142352: F = 23.6937: D = .000105244#
IF GA = 7 THEN E = .152593: F = 35.1717: D = .000057679#
IF GA = 8 THEN E = .149441: F = 29.379: D = .000101154#

CLS
GOTO 400:

300 LOCATE 4, 13: INPUT "G-Function; Siacci = 1, Russian G43 = 3"; GS
TE = 289.08: Pr = 750: Tc = 15
IF GS = 1 THEN E = .362013#: F = 93.1823: D = .000212
IF GS = 3 THEN E = .157713: F = 36.39542: D = .00007454#
CLS
GOTO 400:

400 LOCATE 5, 12: INPUT "Would You Like to Cancel the data File [Y/N]"; y$
IF y$ = "y" OR y$ = "Y" THEN KILL "c:\koef.dat"
CLS
```

```
LOCATE 5, 12: INPUT "Guess Initial Coefficient = "; koef
22 LOCATE 7, 12: INPUT "Projectile Speed = "; speed
LOCATE 8, 12: INPUT "Departure Angle = "; kk
LOCATE 9, 12: INPUT "x- coordinate of Target [m] = "; dis
LOCATE 10, 12: INPUT "Y - coordinate of Target over Firing Site = "; yy
LOCATE 11, 12: INPUT "y - coordinate of Gun over Firing Site [m] = "; vo
LOCATE 13, 12: INPUT "Error in x-coordinate = "; gab
LOCATE 14, 12: PRINT "Number of Steps when range Ends with 0 is [10]"
LOCATE 15, 12: PRINT "Number of Steps when range Ends with 00 is [100]"
LOCATE 16, 12: PRINT "Number of Steps range Ends with a NON ZERO Number is [1]"
LOCATE 17, 12: INPUT "Number of Steps = "; n
LOCATE 19, 13: INPUT "Temperature of Dry Air at Firing Site [C] = "; ta
LOCATE 20, 13: INPUT "Temperature of Propellant [C] = "; tp
LOCATE 21, 13: INPUT "Atmospheric Pressure at firing site [mmHg] = "; pa
LOCATE 22, 13: INPUT "Humidity at Firing Site [decimal #] = "; ea
ta = ta + 273.15
IF ta > 273.16 AND ta <= 327.15 THEN
    ea = ea * 7.50187 * EXP(19.04 * (1 - 280.07 / ta))
END IF
IF ta > 255.15 AND ta < 273.15 THEN
    ea = ea * 7.50187 * EXP(22.024 * (1 - 279.24 / ta))
END IF


pa1 = ta / (1 - .3785 * ea / pa)
speed = (speed + .001 * speed * (tp - Tc))
CLS
END SUB

SUB KthimiKendit (kk)
kk = kk
END SUB
SUB menu (cog, cof, xf, yf, xfu, yfu, t$)
COLOR cog, cof
LOCATE xf - 1, yf: PRINT t$
LOCATE xf, yf: PRINT "É" + STRING$(yfu - yf, 205) + "»";
FOR i = xf + 1 TO xfu
   LOCATE i, yf: PRINT "º" + SPACE$(yfu - yf) + "º";
NEXT
LOCATE xfu + 1, yf: PRINT "È" + STRING$(yfu - yf, 205) + "¼";
END SUB
SUB NPkoef (k, l, r, Q, h, y1, z1, v1, w1)
k = h * y1: l = h * z1
r = h * v1: Q = h * w1
END SUB
```

```
SUB NPxyzvw (nk, x, x0, y, y0, z, z0, v, v0, w, w0, h, h0, k, l, r, Q)
IF nk = 1 THEN
   x = x0: y = y0: z = z0
   v = v0: w = w0: h = h0
   GOTO fund:
END IF

IF nk = 2 OR nk = 3 THEN
   x = x0 + (.5 * h): y = y0 + (.5 * k)
   z = z0 + (.5 * l): v = v0 + (.5 * r)
   w = w0 + (.5 * Q)
   GOTO fund:
END IF
IF nk = 4 THEN
   x = x0 + h: y = y0 + k: z = z0 + l
   v = v0 + r: w = w0 + Q
END IF
fund:
END SUB
SUB Rezervim (koef, kek, gab, x0)
OPEN "c:\koef.dat" FOR APPEND AS #1
PRINT #1, koef, kek, gab, x0
CLOSE #1
END SUB

SUB y1z1v1w1 (x, y, z, v, w, y1, z1, v1, w1, koef, E, F, D, TE, pa1, bb, ta1, Pr, pa)

bb = y * SQR(1 + z ^ 2)
IF bb > 256! THEN
   y1 =-1 * koef* (pa / Pr) * (TE / pa1) ^ .5 * ((pa1 - .006328 * v) / pa1) ^ 4.4 * (E * bb - F) / bb
ELSE
   y1 =-1 * koef* (pa / Pr) * (TE / pa1) ^ .5 * ((pa1 - .006328 * v) / pa1) ^ 4.4 * D * (bb) ^ 2 / bb
END IF
z1 = -9.80665 / y ^ 2
v1 = z
w1 = 1 / y
END SUB
```

APPENDIX D

PC QuickBasic Program
APROJ16.BAS
Reference G-functions: G1, G2, ... G7, etc.

'
'
'
'
'   Ballistics Coefficient is Known
'   Non Standard Atmosphere, Wind Present

'Find: Departure Angle, Time of Flight, Impact Speed, Impact Angle, etc.
'Given: The coordinates of the target and the location of the muzzle of the cannon
'   & Ballistics Coefficient
'   Temperature of air and thrusting charge of projectile are known;
'   The weight of projectile and the air humidity are known
'   Projectile flight is in presence of wind.
'---
'   CONTROL DATA
' Example 1   G7 function, Horizontal Shooting
'   Select ICAO atmosphere. No wind
'   Select G7 = 7
'   Select: Horizontal Shooting (Press 3)

'INPUT: x0 = 1500, projectile speed = 830
'   Temperature of air = 15,Temperature of propellant= 21.11
'   Presure = 760, Humidity of Air = 0.00, projectile mass = 1, change in mass = 0
'   Range wind = 0.0, cross wind, 0.00, BC = 3.772

221

'RESULTS:Launching Angle = 1.1685 degree, Time of Flight = 2.86s
'   Impact Speed = 346.33 m/s, Impact Angle = -2.098 Degree
'   Coordinates of the trajectory vertex (863.7, 10.2), wind deflection = 0m

'Example 2 Inclined Shooting
'   Select ICAO atmosphere, Wind present
'   Select G7 = 7
'   Select: Inclined Shooting (Press 1)

'INPUT: Inclined Distance = 1500, Elevation Angle = 30, projectile velocity = 830
'   y-coordinate of Muzzle = 0, Temperature of air = 15,Temperature of propellant= 21.11
'   Presure = 760, Humidity of Air = 0.00, projectile mass = 1, change in mass = 0
'   range wind = 4.5, cross wind = 5, BC = 3.772

'RESULTS: Launching Angle = 30.8552 degree, (super elevation angle = 0.8552 degreee),
'   Time of Flight = 2.827 s, Impact Speed = 348.88 m/s, Impact Angle = 28.219 Degree
'   Coordinates of the trajectory vertex (0, 0): No Vertex, wind deflection = 5.02m
----
'Functions and Sub. Prog.

DECLARE SUB y1z1v1w1 (x, y, z, v, w, y1, z1, v1, w1, koef, pa1, wind, ys, yy, pa, ta1, TE, Pr, E, F, D)
DECLARE SUB InfHyres (xx, voo, vo1, ta, ta1, pa, pa1, ea, m, dm, tp, wind, xc, yc, vo, yc1, cw, xx1, koef, atm, G, GA, GS, TE, Pr, E, F, D, Tc, xT, yT)
DECLARE SUB InfDales (x0, y0, z0, v0, w0, xc, yc, A, aT, vo, yc1, xm, ym, cw, xx, voo, xa, ya, ta, va, aa)
DECLARE SUB NPxyzvw (nk, x, x0, y, y0, z, z0, v, v0, w, w0, h, h0, k, l, r, q)
DECLARE SUB NPkoef (k, l, r, q, h, y1, z1, v1, w1)
DECLARE SUB menu (cog, cof, xf, yf, xfu, yfu, t$)
DECLARE SUB y0z0 (y0, z0, A, vo1, wind)
DECLARE SUB c (koef)
'Variables
SCREEN 0
1:
DIM m(4, 4), v(4) 'Intermediate values (k,l,r,q)
rendi = 4 'rend dif.
cog = 7: cof = 0
cikli = 0
A = 23 'Initial Angle 23 degree
kendi = 22 'Angle [Degree] for maximum distance
kov = 1 'Test of the value of v0
gab = 0.1 'error 0.1 m.

ELEMENTS OF EXTERIOR BALLISTICS

```
tt = 1
Incl = 0
'Solution
CLS

'Initial Data
menu cog, cof, 3, 10, 7, 70, "DATA INPUT"
InfHyres xx, voo, vo1, ta, ta1, pa, pa1, ea, m, dm, tp, wind, xc, yc, yc1, vo, cw, xx1, koef,
atm, G, GA, GS, TE, Pr, E, F, D, Tc, Incl, El
hap = 0.1
'Initial values
F:
x0 = 0: v0 = vo: w0 = 0
y0z0 y0, z0, A, vo1, wind: h0 = hap
c koef
ff:
FOR nk = 1 TO rendi
    NPxyzvw nk, x, x0, Y, y0, z, z0, v, v0, w, w0, h, h0, k, l, r, q
    y1z1v1w1 x, Y, z, v, w, y1, z1, v1, w1, koef, pa1, wind, ys, yy, pa, ta1, TE, Pr, E, F, D
    NPkoef k, l, r, q, h, y1, z1, v1, w1
    m(nk, 1) = k: m(nk, 2) = l
    m(nk, 3) = r: m(nk, 4) = q
NEXT nk

'Estimation for new points
FOR i = 1 TO rendi
    v(i) = 1 / 6 * (m(1, i) + 2 * m(2, i) + 2 * m(3, i) + m(4, i))
NEXT i

'New Points
x0 = x0 + h: y0 = y0 + v(1): z0 = z0 + v(2)
v0 = v0 + v(3): w0 = w0 + v(4)
    ymm = v0 + wind * z0 * w0
    xmm = x0 + wind * w0
IF ABS(z0) <= .0001 THEN
    xm = xmm
    ym = ymm
END IF

'Tests the y-value
IF Incl = 1 THEN
    yT = v0 + wind * z0 * w0
ELSE
    yT = v0
END IF
xT = x0 + wind * w0
```

```
IF kov = 1 THEN kov = -1: GOTO ff:
IF ABS(xT - xc) < gab AND ABS(yT - vo - yc) <= (gab * TAN(A * 3.1415954# / 180)) THEN
   c:
   'DISPLAY of RESULTS
   CLS
   PLAY "a8a16a32b8"
   menu 12, 0, 5, 10, 11, 70, "RESULTS:"
   COLOR 7
   InfDales x0, y0, z0, v0, w0, A, xc, yc, aT, vo, yc1, xm, ym, cw, xx, voo, xa, ya, ta, va, aa, xT, yT
   CLS
   GOTO 1:
END IF
IF ABS(xT - xc) < gab AND (yT - vo - yc) > (gab * TAN(A * 3.1415954# / 180)) THEN
   t$ = " * ? *"
   menu 18, 0, 10, 20, 14, 60, t$
   COLOR 14
   LOCATE 12, 30: PRINT "Wait a moment, Please (+)";
   LOCATE 12, 53: PRINT tt
   tt = tt + 1
   COLOR 7
   A = A - kendi
   GOTO fff:
END IF

IF ABS(xT - xc) < gab AND (yT - vo - yc) < ((-1 * gab) * TAN(A * 3.1415954# / 180)) THEN
   t$ = " * ? *"
   menu 18, 0, 10, 20, 14, 60, t$
   COLOR 14
   LOCATE 12, 30: PRINT "Wait a moment, Please (-)";
   LOCATE 12, 53: PRINT tt
   tt = tt + 1
   COLOR 7
   A = A + kendi
   GOTO fff:
END IF
GOTO ff:
fff:
'Restart Cycle
cikli = cikli + 1
IF cikli = 20 THEN GOTO c:
kendi = kendi / 2
kov = 1
GOTO F:
SUB c (koef)
koef = koef
```

END SUB

SUB InfDales (x0, y0, z0, v0, w0, A, xc, aT, yc, vo, yc1, xm, ym, cw, xx, voo, xa, ya, ta, va, aa, xT, yT)
aT = ATN(z0) * 180 / 3.141592654#
LOCATE 6, 16: PRINT "Departure Angle   :"; A; "Degree"
LOCATE 7, 16: PRINT "Time of Flight to Target :"; INT((w0) * 1000 + .5) / 1000
LOCATE 8, 16: PRINT "Terminal Speed   :"; INT((y0 / COS(ATN(z0))) * 100 + .5) / 100; " m/s"
LOCATE 9, 16: PRINT "Terminal Angle   :"; aT; " degree"
LOCATE 10, 16: PRINT "Coordinates of Vertex :"; "("; INT((xm) * 10 + .5) / 10; ","; INT((ym) * 10 + .5) / 10; ")"
LOCATE 11, 16: PRINT "Cross-Wind Deflection :"; INT((cw * (w0 - xx / (voo * COS(A * 3.14159265# / 180)))) * 100 + .5) / 100
LOCATE 13, 16: PRINT "Location of TARGET   :"; "("; ((xT) * 10 + 0.5) / 10; ((yT - vo) * 100 + 0.5) / 100; ")"
LOCATE 14, 16: PRINT "Location of FIREARM   :"; "(0,"; vo; ")"
LOCATE 15, 16: PRINT "Distance to TARGET   :"; (((xT ^ 2 + (yT - 2 * vo) ^ 2) ^ .5) * 10 + 0.5) / 10
COLOR 7
LOCATE 24, 11: PRINT " Pres [P] to repeat [Esc] to end ";
cc$ = INPUT$(1)
IF cc$ = CHR$(27) THEN SCREEN 9: CLS: END
END SUB

SUB InfHyres (xx, voo, vo1, ta, ta1, pa, pa1, ea, m, dm, tp, wind, xc, yc, yc1, vo, cw, xx1, koef, atm, G, GA, GS, TE, Pr, E, F, D, Tc, Incl, El)

LOCATE 4, 15: INPUT "ATMOSPHERE: ASM = 1, ICAO = 2; TSA = 3 "; atm
CLS
IF atm = 1 THEN GOTO 100:
IF atm = 2 THEN GOTO 200:
IF atm = 3 THEN GOTO 300:

100 LOCATE 4, 2: INPUT "G-Function; G1 = 1, G2 = 2, G5 = 5, G6 = 6, G7 = 7, G8 = 8, Siacci=9, G43=43"; G
TE = 289.6: Pr = 750: Tc = 21.11
IF G = 1 THEN E = .312914: F = 79.3976: D = .000100347#
IF G = 2 THEN E = .1413: F = 29.9097: D = .00009116233#
IF G = 5 THEN E = .19734: F = 49.0806: D = .0000743571#
IF G = 6 THEN E = .140533: F = 23.6633: D = .00010334#
IF G = 7 THEN E = .150355: F = 34.7319: D = .000056648#
IF G = 8 THEN E = .14733: F = 29.085: D = .000099366#
IF G = 9 THEN E = .36223: F = 93.5778: D = 0.00012078
IF G = 43 THEN E = 0.163211: F = 38.7839: D = 0.000074482
CLS

GOTO 400:

200 LOCATE 4, 2: INPUT "G-Function; G1 = 1, G2 = 2, G5 =5, G6 = 6, G7 = 7, G8=8, Siacci=9, G43=43"; GA
TE = 288.15: Pr = 760: Tc = 21.11
IF GA = 1 THEN E = .315754: F = 78.6769: D = .00010584#
IF GA = 2 THEN E = .143353: F = 30.2415: D = .00009287#
IF GA = 5 THEN E = .200207: F = 49.625: D = .000075244#
IF GA = 6 THEN E = .142352: F = 23.6937: D = .000105244#
IF GA = 7 THEN E = .152593: F = 35.1717: D = .000057679#
IF GA = 8 THEN E = .149441: F = 29.379: D = .000101154#
IF GA = 9 THEN E = 0.36874: F = 95.2603: D = 0.00012296
IF GA = 43 THEN E = 0.165114: F = 39.2361: D = 0.00007535
CLS
GOTO 400:

300 LOCATE 4, 13: INPUT "G-Function: Siacci = 1, Russian G43 = 3"; GS
TE = 289.08: Pr = 750: Tc = 15
IF GS = 1 THEN E = .362013: F = 93.1823: D = .00012094
IF GS = 3 THEN E = .157713: F = 36.39542: D = .00007454#
CLS
GOTO 400:

400
LOCATE 3, 12: INPUT "Distance: Inclined = 1, Declined =2, Horizontal = 3 "; Incl
IF Incl = 1 THEN
   LOCATE 4, 16: INPUT "Inclined Distance [m]   = "; xx
   LOCATE 5, 16: INPUT "Elevation Angle   = "; El
END IF
IF Incl = 2 THEN
   LOCATE 4, 15: INPUT "Declined Distance [m]   = "; xx
   LOCATE 5, 16: INPUT "Depression Angle   = "; El
   xx = xx * COS(El * 3.141592654# / 180)
   El = 0
END IF
IF Incl = 3 THEN
   LOCATE 4, 15: INPUT "Horizontal Distance [m]   = "; xx
   El = 0
END IF
CLS

LOCATE 5, 13: INPUT "y-Coordinate of Muzzle [m]   = "; vo
yc = xx * SIN(El * 3.141592654# / 180)
yc1 = yc
xx = xx * COS(El * 3.141592654# / 180)
xc = xx

ELEMENTS OF EXTERIOR BALLISTICS

```
LOCATE 6, 13: INPUT "Departure Velocity [m/s]    = "; voo
LOCATE 7, 13: INPUT "Temperature of Air [C]      = "; ta
LOCATE 8, 13: INPUT "Propellant Temperature [C]  = "; tp
LOCATE 9, 13: INPUT "Atmospheric Pressure [mm Hg] = "; pa
LOCATE 10, 13: INPUT "Hunidity of Air [Decimal]  = "; ea
LOCATE 11, 13: INPUT "Projectile Standard Mass [kg] = "; m
LOCATE 12, 13: INPUT "Change in Projectile Mass  = "; dm
LOCATE 13, 13: INPUT "Range-Wind [m/s]           = "; wind
LOCATE 14, 13: INPUT "Cross-Wind [m/s]           = "; cw
LOCATE 15, 13: INPUT "Ballistics Coefficient BC  = "; koef
CLS

ta = ta + 273.15
IF ta > 273.16 AND ta <= 327.15 THEN
    ea = ea * 7.50187 * EXP(19.04 * (1 - 280.07 / ta))
END IF
IF ta > 255.15 AND ta < 273.15 THEN
    ea = ea * 7.50187 * EXP(22.024 * (1 - 279.24 / ta))
END IF

pa1 = ta / (1 - .3785 * ea / pa)
vo1 = (voo - .4 * voo * (dm / m) + .001 * voo * (tp - Tc))
koef = koef * (1 - dm / m)
END SUB

SUB menu (cog, cof, xf, yf, xfu, yfu, t$)
COLOR cog, cof
LOCATE xf - 1, yf: PRINT t$
LOCATE xf, yf: PRINT "É" + STRING$(yfu - yf, 205) + "»";
FOR i = xf + 1 TO xfu
    LOCATE i, yf: PRINT "º" + SPACE$(yfu - yf) + "º";
NEXT
LOCATE xfu + 1, yf: PRINT "È" + STRING$(yfu - yf, 205) + "¼";
END SUB

SUB NPkoef (k, l, r, q, h, y1, z1, v1, w1)
k = h * y1: l = h * z1
r = h * v1: q = h * w1
END SUB

SUB NPxyzvw (nk, x, x0, y, y0, z, z0, v, v0, w, w0, h, h0, k, l, r, q)
IF nk = 1 THEN
    x = x0: y = y0: z = z0
    v = v0: w = w0: h = h0
    GOTO fund:
END IF
```

```
IF nk = 2 OR nk = 3 THEN
   x = x0 + (.5 * h): y = y0 + (.5 * k)
   z = z0 + (.5 * l): v = v0 + (.5 * r)
   w = w0 + (.5 * q)
   GOTO fund:
END IF
IF nk = 4 THEN
   x = x0 + h: y = y0 + k: z = z0 + l
   v = v0 + r: w = w0 + q
END IF
fund:
END SUB

SUB y0z0 (y0, z0, A, vo1, wind)
y0 = SQR(vo1 ^ 2 + wind ^ 2 - 2 * vo1 * wind * COS(A * 3.141592654# / 180))
y0 = y0 * COS(A * 3.141592654# / 180)
z0 = TAN(A * 3.141592654# / 180)
z0 = z0 / (1 - wind / (vo1 * COS(A * 3.141592654# / 180)))
END SUB

SUB y1z1v1w1 (x, y, z, v, w, y1, z1, v1, w1, koef, pa1, wind, ys, yy, pa, ta1, TE, Pr, E, F, D)
ta1 = (TE / pa1) ^ .5
yy = y * SQR(1 + z ^ 2)
IF yy > 271! THEN
    y1 = -1 * koef * (pa / Pr) * ta1 * ((pa1 - .006328 * v) / pa1) ^ 4.4 * (E * yy - F) / yy
ELSE
    y1 = -1 * koef * (pa / Pr) * ta1 * ((pa1 - .006328 * v) / pa1) ^ 4.4 * D * (yy) ^ 2 / yy
END IF
z1 = -9.80665 / y ^ 2
v1 = z
w1 = 1 / y
END SUB
```

ELEMENTS OF EXTERIOR BALLISTICS | 229

PC QuickBasic Program
Alapua16.BAS
Lapua G-function of bullet 0.338, 300Gr., GB528 (BC=3.796)
Ballistics Coefficient is Known
Non-Standard Atmosphere, Wind Present

'Find: Departure Angle, Time of Flight, Impact Speed, Impact Angle, etc.
'Given: The coordinates of the target and the location of the muzzle of the rifle
' & Ballistics Coefficient
' Temperature of air and temperature of propellant charge of the projectile are known;
' The mass of projectile and the air humidity are known
' Projectile flight is in presence of wind.
'--

' CONTROL DATA
'EXAMPLE 1: UPHILL Shooting
' Select ICAO Atmosphere, GB528 function, Select Inclined Shooting
'INPUT: ICAO atmosphere (press 2) , function of resistance GB528 Lapua bullet (press 5),
'
' Inclined Range = 1500, Elevation Angle = 30, y-coordinate of Muzzle = 0,
' Projectile velocity = 830, Temperature of air = 15 , Temperature of propellant= 21.11
' Presure = 760 mmHg, Humidity of Air = 0, projectile mass = 0.01944 kg/m3 (300 Gr)
' (or any number, mass = 1) change in mass = 0
' range wind = 4.5, cross wind, 0, BC = 3.796

'RESULTS: Departure Angle = 30.831 degree, Time of Flight = 2.792s,
' Impact Speed = 352.25 m/s, Impact Angle = 28.268
' Coordinates of the trajectory vertex (0, 0), cross-wind deflection = 0.
'
'EXAMPLE 2: DOWNHILL Shooting
' Select ICAO Atmosphere, GB528 function, Select Declined Range
'INPUT: ICAO atmosphere (press 2) , function of resistance GB528 Lapua bullet (press 5),
'
' Declined Range = 1500, Depression Angle = 30, y-coordinate of Muzzle = 750,
' Projectile velocity = 830, Temperature of air = 15 , Temperature of propellant= 21.11
' Presure = 760 mmHg, Humidity of Air = 0, projectile mass = 0.01944 kg/m3 (300 Gr)
' (or any number, mass = 1) change in mass = 0
' range wind = 0, cross wind, 0, BC = 3.796

'RESULTS: Departure Angle = 0.8533 degree, Time of Flight = 2.213s,

'   Impact Speed = 418.31 m/s, Impact Angle = - 1.349

' Example 3 HORIZONTAL Shooting
'   Horizontal Range = 1500, y-coordinate of Muzzle = 0,
'   Projectile velocity = 830, Temperature of air = 15 , Temperature of propellant= 21.11
'   Presure = 760 mmHg, Humidity of Air = 0, projectile mass = 0.01944 kg/m3 (300 Gr)
'   (or any number, mass = 1) change in mass = 0
'   range wind = 0, cross wind, 0, BC = 3.796

'RESULTS: Departure Angle = 1.1417 degree, Time of Flight = 2.824s,
'   Impact Speed = 349.34 m/s, Impact Angle = - 2.039
'---
'Functions and Sub. Prog.

DECLARE SUB y1z1v1w1 (x, y, z, v, w, y1, z1, v1, w1, koef, pa1, wind, ys, yy, pa, ta1, TE, Pr, E, F, D)
DECLARE SUB InfHyres (xx, voo, vo1, ta, ta1, pa, pa1, ea, m, dm, tp, wind, xc, yc, vo, yc1, cw, xx1, koef, atm, G, GA, GS, TE, Pr, E, F, D, D1, Tc, xT, yT)
DECLARE SUB InfDales (x0, y0, z0, v0, w0, xc, yc, A, aT, vo, yc1, xm, ym, cw, xx, voo, xa, ya, ta, va, aa)
DECLARE SUB NPxyzvw (nk, x, x0, y, y0, z, z0, v, v0, w, w0, h, h0, k, l, r, q)
DECLARE SUB NPkoef (k, l, r, q, h, y1, z1, v1, w1)
DECLARE SUB menu (cog, cof, xf, yf, xfu, yfu, t$)
DECLARE SUB y0z0 (y0, z0, A, vo1, wind)
DECLARE SUB c (koef)
'Variables
SCREEN 0
1:
DIM m(4, 4), v(4) 'Intermediate values (k,l,r,q)
rendi = 4 'rend dif.
cog = 7: cof = 0
cikli = 0
A = 23 'Initial Angle 23 degree
kendi = 22 'Angle [Degree] for maximum distance
kov = 1 'Test of the value of v0
gab = 0.1 'error 0.1 m.
tt = 1
Incl = 0
'Solution
CLS

'Initial Data
menu cog, cof, 3, 10, 7, 70, "DATA INPUT"

InfHyres xx, voo, vo1, ta, ta1, pa, pa1, ea, m, dm, tp, wind, xc, yc, yc1, vo, cw, xx1, koef, atm, G, GA, GS, TE, Pr, E, F, D, D1, Tc, Incl, El
hap = 0.1
'Initial values
F:
x0 = 0: v0 = vo: w0 = 0
y0z0 y0, z0, A, vo1, wind: h0 = hap
c koef
ff:
FOR nk = 1 TO rendi
   NPxyzvw nk, x, x0, Y, y0, z, z0, v, v0, w, w0, h, h0, k, l, r, q
   y1z1v1w1 x, Y, z, v, w, y1, z1, v1, w1, koef, pa1, wind, ys, yy, pa, ta1, TE, Pr, E, F, D, D1
   NPkoef k, l, r, q, h, y1, z1, v1, w1
   m(nk, 1) = k: m(nk, 2) = l
   m(nk, 3) = r: m(nk, 4) = q
NEXT nk
'Estimation for new points
FOR i = 1 TO rendi
   v(i) = 1 / 6 * (m(1, i) + 2 * m(2, i) + 2 * m(3, i) + m(4, i))
NEXT i

'New Points
x0 = x0 + h: y0 = y0 + v(1): z0 = z0 + v(2)
v0 = v0 + v(3): w0 = w0 + v(4)
ymm = v0 + wind * z0 * w0
xmm = x0 + wind * w0
IF ABS(z0) <= .0001 THEN
   xm = xmm
   ym = ymm
END IF
'Tests the y-value
IF Incl = 1 THEN
   yT = v0 + wind * z0 * w0
ELSE
   yT = v0
END IF
xT = x0 + wind * w0
IF kov = 1 THEN kov = -1: GOTO ff:

IF ABS(xT - xc) < gab AND ABS(yT - vo - yc) <= (gab * TAN(A * 3.1415954# / 180)) THEN
  c:
  'DISPLAY of RESULTS
  CLS
  PLAY "a8a16a32b8"
  menu 12, 0, 5, 10, 11, 70, "RESULTS:"
  COLOR 7

InfDales x0, y0, z0, v0, w0, A, xc, yc, aT, vo, yc1, xm, ym, cw, xx, voo, xa, ya, ta, va, aa, xT, yT
CLS
GOTO 1:
END IF
IF ABS(xT - xc) < gab AND (yT - vo - yc) > (gab * TAN(A * 3.1415954# / 180)) THEN
  t$ = " * ? *"
  menu 18, 0, 10, 20, 14, 60, t$
  COLOR 14
  LOCATE 12, 30: PRINT "Wait a moment, Please (+)";
  LOCATE 12, 53: PRINT tt
  tt = tt + 1
  COLOR 7
  A = A - kendi
  GOTO fff:
END IF

IF ABS(xT - xc) < gab AND (yT - vo - yc) < ((-1 * gab) * TAN(A * 3.1415954# / 180)) THEN
  t$ = " * ? *"
  menu 18, 0, 10, 20, 14, 60, t$
  COLOR 14
  LOCATE 12, 30: PRINT "Wait a moment, Please (-)";
  LOCATE 12, 53: PRINT tt
  tt = tt + 1
  COLOR 7
  A = A + kendi
  GOTO fff:
END IF
GOTO ff:

fff:
'Restart Cycle
cikli = cikli + 1
IF cikli = 20 THEN GOTO c:
kendi = kendi / 2
kov = 1
GOTO F:
SUB c (koef)
koef = koef
END SUB

SUB InfDales (x0, y0, z0, v0, w0, A, xc, aT, yc, vo, yc1, xm, ym, cw, xx, voo, xa, ya, ta, va, aa, xT, yT)
aT = ATN(z0) * 180 / 3.141592654#
LOCATE 6, 16: PRINT "Departure Angle :"; A; "Degree"
LOCATE 7, 16: PRINT "Time of Flight to Target :"; INT((w0) * 1000 + .5) / 1000

ELEMENTS OF EXTERIOR BALLISTICS

LOCATE 8, 16: PRINT "Terminal Speed :"; INT((y0 / COS(ATN(z0))) * 100 + .5) / 100; " m/s"
LOCATE 9, 16: PRINT "Terminal Angle :"; aT; " degree"
LOCATE 10, 16: PRINT "Coordinates of Vertex :"; "("; INT((xm) * 10 + .5) / 10; ",";
INT((ym) * 10 + .5) / 10; ")"
LOCATE 11, 16: PRINT "Cross-Wind Deflection :"; INT((cw * (w0 - xx / (voo * COS(A * 3.14159265# / 180)))) * 100 + .5) / 100
LOCATE 13, 16: PRINT "Location of TARGET :"; "("; ((xT) * 10 + 0.5) / 10; ((yT) * 100 + 0.5) / 100; ")"
LOCATE 14, 16: PRINT "Location of FIREARM :"; "(0,"; vo; ")"
LOCATE 15, 16: PRINT "Distance to TARGET :"; (((xT ^ 2 + (yT - vo) ^ 2) ^ .5) * 10 + 0.5) / 10
COLOR 7
LOCATE 24, 11: PRINT " Pres [P] to repeat [Esc] to end ";
cc$ = INPUT$(1)
IF cc$ = CHR$(27) THEN SCREEN 9: CLS: END
END SUB

SUB InfHyres (xx, voo, vo1, ta, ta1, pa, pa1, ea, m, dm, tp, wind, xc, yc, yc1, vo, cw, xx1, koef, atm, G, GA, GS, TE, Pr, E, F, D, D1, Tc, Incl, El)
LOCATE 4, 16: INPUT "Atmosphere: ASM = 1, ICAO = 2"; atm
CLS
IF atm = 1 THEN GOTO 100:
IF atm = 2 THEN GOTO 200:

100 LOCATE 4, 16: INPUT "G-Function: GB528 LAPUA = 5 "; G
TE = 289.6: Pr = 750: Tc = 21.11
IF G = 5 THEN E = 0.13895: F = 29.709: D = .0242: D1 = 1.399
CLS
GOTO 400:

200 LOCATE 4, 16: INPUT "G-Function: GB528 Lapua = 5"; GA
TE = 288.15: Pr = 760: Tc = 21.11
IF GA = 5 THEN E = 0.141: F = 30.031: D = .02438: D1 = 1.3696
CLS
GOTO 400:

400
LOCATE 3, 12: INPUT "Distance: Inclined = 1, Declined =2, Horizontal = 3 "; Incl
IF Incl = 1 THEN
   LOCATE 4, 16: INPUT "Inclined Distance [m]   = "; xx
   LOCATE 5, 16: INPUT "Elevation Angle   = "; El
END IF
IF Incl = 2 THEN
   LOCATE 4, 15: INPUT "Declined Distance [m]   = "; xx
   LOCATE 5, 16: INPUT "Depression Angle   = "; El

```
        xx = xx * COS(El * 3.141592654# / 180)
        El = 0
    END IF
    IF Incl = 3 THEN
        LOCATE 4, 15: INPUT "Horizontal Distance [m]        = "; xx
        El = 0
    END IF
    CLS

    LOCATE 5, 13: INPUT "y-Coordinate of Muzzle [m]    = "; vo
    yc = xx * SIN(El * 3.141592654# / 180)
    yc1 = yc
    xx = xx * COS(El * 3.141592654# / 180)
    xc = xx
    LOCATE 6, 13: INPUT "Departure Velocity [m/s]      = "; voo
    LOCATE 7, 13: INPUT "Temperature of Air [C]        = "; ta
    LOCATE 8, 13: INPUT "Propellant Temperature [C]    = "; tp
    LOCATE 9, 13: INPUT "Atmospheric Pressure [mm Hg]  = "; pa
    LOCATE 10, 13: INPUT "Hunidity of Air [Decimal]    = "; ea
    LOCATE 11, 13: INPUT "Projectile Standard Mass [kg] = "; m
    LOCATE 12, 13: INPUT "Change in Projectile Mass    = "; dm
    LOCATE 13, 13: INPUT "Range-Wind [m/s]             = "; wind
    LOCATE 14, 13: INPUT "Cross-Wind [m/s]             = "; cw
    LOCATE 15, 13: INPUT "Ballistics Coefficient BC    = "; koef
    CLS
    ta = ta + 273.15
    IF ta > 273.16 AND ta <= 327.15 THEN
        ea = ea * 7.50187 * EXP(19.04 * (1 - 280.07 / ta))
    END IF

    IF ta > 255.15 AND ta < 273.15 THEN
        ea = ea * 7.50187 * EXP(22.024 * (1 - 279.24 / ta))
    END IF

    pa1 = ta / (1 - .3785 * ea / pa)
    vo1 = (voo - .4 * voo * (dm / m) + .001 * voo * (tp - Tc))
    koef = koef * (1 - dm / m)
END SUB

SUB menu (cog, cof, xf, yf, xfu, yfu, t$)
    COLOR cog, cof
    LOCATE xf - 1, yf: PRINT t$
    LOCATE xf, yf: PRINT "É" + STRING$(yfu - yf, 205) + "»";

    FOR i = xf + 1 TO xfu
        LOCATE i, yf: PRINT "º" + SPACE$(yfu - yf) + "º";
```

```
NEXT
LOCATE xfu + 1, yf: PRINT "È" + STRING$(yfu - yf, 205) + "¼";
END SUB

SUB NPkoef (k, l, r, q, h, y1, z1, v1, w1)
k = h * y1: l = h * z1
r = h * v1: q = h * w1
END SUB

SUB NPxyzvw (nk, x, x0, y, y0, z, z0, v, v0, w, w0, h, h0, k, l, r, q)

IF nk = 1 THEN
  x = x0: y = y0: z = z0
  v = v0: w = w0: h = h0
  GOTO fund:
END IF

IF nk = 2 OR nk = 3 THEN
  x = x0 + (.5 * h): y = y0 + (.5 * k)
  z = z0 + (.5 * l): v = v0 + (.5 * r)
  w = w0 + (.5 * q)
  GOTO fund:
END IF

IF nk = 4 THEN
  x = x0 + h: y = y0 + k: z = z0 + l
  v = v0 + r: w = w0 + q
END IF

fund:
END SUB

SUB y0z0 (y0, z0, A, vo1, wind)
y0 = SQR(vo1 ^ 2 + wind ^ 2 - 2 * vo1 * wind * COS(A * 3.141592654# / 180))
y0 = y0 * COS(A * 3.141592654# / 180)
z0 = TAN(A * 3.141592654# / 180)
z0 = z0 / (1 - wind / (vo1 * COS(A * 3.141592654# / 180)))
END SUB

SUB y1z1v1w1 (x, y, z, v, w, y1, z1, v1, w1, koef, pa1, wind, ys, yy, pa, ta1, TE, Pr, E, F, D, D1)
ta1 = (TE / pa1) ^ .5
yy = y * SQR(1 + z ^ 2)
IF yy > 325 THEN
  y1 = -1 * koef * (pa / Pr) * ta1 * ((pa1 - .006328 * v) / pa1) ^ 4.4 * (E * yy - F) / yy
ELSE
  y1 = -1 * koef * (pa / Pr) * ta1 * ((pa1 - .006328 * v) / pa1) ^ 4.4 * (D * yy - D1) / yy
```

```
END IF
z1 = -9.80665 / y ^ 2
v1 = z
w1 = 1 / y
END SUB
```

ELEMENTS OF EXTERIOR BALLISTICS | 237

PC QuickBasic Program
ACHA16.BAS
Ballistics Coefficient is Known
Non Standard Atmosphere, Wind Present

    LIST OF BULLETS
    ICAO Atmosphere Characteristic G-functions
  1. Bullet 0.300 Winchester Magnum, Velocity 884m/s, BC = 4.97096: WM = 1 (Press 1)

  2. Caliber 0.308, 168 grain, Sierra International Bullet, Velocity 792.48m/s, BC = 5.6325: SI = 2 (press 2)

  3. M118LR Bullet, 0.308" mass 0.01134 kg. Velocity 884m/s, BC = 5.3970 M118LR = 5 (Press 5)

  4. M118 Ball Bullet (Federal GM308M2), 0.308", mass = 0.00134kg, velocity 792.48m/s, BC =5.3973, ' 'M118BB = 6 (Press 6)

  5. 300 gr. .338 - .416 Bullet (mass 0.01944kg, Velocity 927.40m/s, BC = 3.791 300Gr = 7 (Press 7)

  6. caliber 0.30 Ball M2 Bullet, mass 0.00972 kg, Velocity 853.44 m/s, BC =6.2965 BM2 = 8 (Press 8)

'Find: Departure Angle, Time of Flight, Impact Speed, Impact Angle, etc.

'Given: The coordinates of the target and the location of the muzzle of the firearm are known
    initial velocity and Ballistics Coefficient are known
    Temperature of air, pressure and humidity are known;
    Projectile flight is in presence of wind.(range-wind and cross-wind are known

'NOTE. The program asks for inclined shootings or Horizontal shooting.
    Input 1 for inclined shooting, and 2 or any number for Horizontal shooting
'---
    CONTROL DATA
'EXAMPLE 1 Horizontal Shooting, Sierra International Bullet: SI = 2
    Select ASM = 1
    Select Horizontal Shooting: Press 3

'INPUT: Range = 548.64m (600yards), y-coordinate of muzzle = 0, projectile speed = 792.48
    Temperature of air = 15,Temperature of propellant = 21.11

'   Presure = 750, Humidity of Air = 0.78, projectile mass = 1 (any number different from zero) when change in mass = 0
'   Range wind = 0, cross wind, 0, BC = 5.6325

'RESULTS:Launching Angle = 0.3491 Degree, Time of Flight = 0.895s
'   Impact Speed = 479.69 m/s, Impact Angle = -.488 Degree
'   Coordinates of the trajectory vertex (301, 1), wind deflection 0.0 m

' Example 2 Downhill Shooting. Bullet 0.300 Winchester Magnum, Velocity 884m/s, BC = 4.9709. ICAOP Atmosphere
'   Declined range = 1000m, Depression angle = 30 degree. Coordinates of muzzzle (0, 500)
'   Coordinates of target =(866.03, 0). No Wind.
'
' Input: WM = 1 (Press 1), Declined Range (Press 2), Declined Range = 1000, Depression Angle = 30,
'   y-coordinate of muzzle = 500, Initial Velocity = 884, temperature 15, Propellant Temperature = 21.11
'   Pressure = 760, Humidity = 0, projectile mass = 1, BC = 4.9709
'
'Results: Departure Angle = 0.5164, time = 1.412s, velocity 432.45, impact angle = -0.833, Vertex (488.8, 502.5).

'Example 3: Consider the data in Example 2 and a Range wind = 5m/s, and a Cross-wind of 5m/s.

'Results: Departure Angle = 0.5134, time = 1.4052s, velocity 432.07, impact angle = - 0.830, Vertex (487.8, 502.4).
'   Cross-wind deflection = 2.13m

' Example 4: Consider the data in Example 2, but instead of downhill shooting consider up-hill shooting.
'   Coordinates of muzzle (0, 0)

' Results: Departure angle = 30.5894 degree (super-elevation angle = 0.5894 degree), time = 1.781s, velocity = 371.95,
'   impact angle = 28.94.

'--
'Functions and Sub. Prog.

DECLARE SUB y1z1v1w1 (x, y, z, v, w, y1, z1, v1, w1, koef, pa1, wind, ys, yy, pa, ta1, TE, Pr, E, F, D)
DECLARE SUB InfHyres (xx, voo, vo1, ta, ta1, pa, pa1, ea, m, dm, tp, wind, xc, yc, vo, yc1, cw, xx1, koef, atm, G, GA, GS, TE, Pr, E, F, D, Tc, xT, yT)

DECLARE SUB InfDales (x0, y0, z0, v0, w0, xc, yc, A, aT, vo, yc1, xm, ym, cw, xx, voo, xa, ya, ta, va, aa)
DECLARE SUB NPxyzvw (nk, x, x0, y, y0, z, z0, v, v0, w, w0, h, h0, k, l, r, q)
DECLARE SUB NPkoef (k, l, r, q, h, y1, z1, v1, w1)
DECLARE SUB menu (cog, cof, xf, yf, xfu, yfu, t$)
DECLARE SUB y0z0 (y0, z0, A, vo1, wind)
DECLARE SUB c (koef)
'Variables
SCREEN 0
1:
DIM m(4, 4), v(4) 'Intermediate values (k,l,r,q)
rendi = 4 'rend dif.
cog = 7: cof = 0
cikli = 0
A = 23 'Initial Angle 23 degree
kendi = 22 'Angle [Degree] for maximum distance
kov = 1 'Test of the value of v0
gab = 0.1 'error 0.1 m.
tt = 1
Incl = 0
'Solution
CLS

'Initial Data
menu cog, cof, 3, 10, 7, 70, "DATA INPUT"
InfHyres xx, voo, vo1, ta, ta1, pa, pa1, ea, m, dm, tp, wind, xc, yc, yc1, vo, cw, xx1, koef, atm, G, GA, GS, TE, Pr, E, F, D, Tc, Incl, El
hap = 0.1
'Initial values
F:
x0 = 0: v0 = vo: w0 = 0
y0z0 y0, z0, A, vo1, wind: h0 = hap
c koef
ff:
FOR nk = 1 TO rendi
    NPxyzvw nk, x, x0, Y, y0, z, z0, v, v0, w, w0, h, h0, k, l, r, q
    y1z1v1w1 x, Y, z, v, w, y1, z1, v1, w1, koef, pa1, wind, ys, yy, pa, ta1, TE, Pr, E, F, D
    NPkoef k, l, r, q, h, y1, z1, v1, w1
    m(nk, 1) = k: m(nk, 2) = l
    m(nk, 3) = r: m(nk, 4) = q
NEXT nk

'Estimation for new points
FOR i = 1 TO rendi
    v(i) = 1 / 6 * (m(1, i) + 2 * m(2, i) + 2 * m(3, i) + m(4, i))
NEXT i

```
'New Points
x0 = x0 + h: y0 = y0 + v(1): z0 = z0 + v(2)
v0 = v0 + v(3): w0 = w0 + v(4)
ymm = v0 + wind * z0 * w0
xmm = x0 + wind * w0
IF ABS(z0) <= .0001 THEN
   xm = xmm
   ym = ymm
END IF
'Tests the y-value
IF Incl = 1 THEN
   yT = v0 + wind * z0 * w0
ELSE
yT = v0
END IF
xT = x0 + wind * w0
IF kov = 1 THEN kov = -1: GOTO ff:
IF ABS(xT - xc) < gab AND ABS(yT - vo - yc) <= (gab * TAN(A * 3.1415954# /
180)) THEN
   c:
   'DISPLAY of RESULTS
   CLS
   PLAY "a8a16a32b8"
   menu 12, 0, 5, 10, 11, 70, "RESULTS:"
   COLOR 7
   InfDales x0, y0, z0, v0, w0, A, xc, yc, aT, vo, yc1, xm, ym, cw, xx, voo, xa, ya, ta, va, aa, xT, yT
   CLS
   GOTO 1:
END IF
IF ABS(xT - xc) < gab AND (yT - vo - yc) > (gab * TAN(A * 3.1415954# / 180)) THEN
   t$ = " * ? *"
   menu 18, 0, 10, 20, 14, 60, t$
   COLOR 14
   LOCATE 12, 30: PRINT "Wait a moment, Please (+)";
   LOCATE 12, 53: PRINT tt
   tt = tt + 1
   COLOR 7
   A = A - kendi
   GOTO fff:
END IF
IF ABS(xT - xc) < gab AND (yT - vo - yc) < ((-1 * gab) * TAN(A * 3.1415954# / 180)) THEN
   t$ = " * ? *"
   menu 18, 0, 10, 20, 14, 60, t$
   COLOR 14
   LOCATE 12, 30: PRINT "Wait a moment, Please (-)";
   LOCATE 12, 53: PRINT tt
```

```
tt = tt + 1
COLOR 7
A = A + kendi
GOTO fff:
END IF
GOTO ff:
fff:
'Restart Cycle
cikli = cikli + 1
IF cikli = 20 THEN GOTO c:
kendi = kendi / 2
kov = 1
GOTO F:
SUB c (koef)
koef = koef
END SUB


SUB InfDales (x0, y0, z0, v0, w0, A, xc, aT, yc, vo, yc1, xm, ym, cw, xx, voo, xa, ya, ta, va, aa, xT, yT)
aT = ATN(z0) * 180 / 3.141592654#
LOCATE 6, 16: PRINT "Departure Angle          :"; A; "Degree"
LOCATE 7, 16: PRINT "Time of Flight to Target :"; INT((w0) * 1000 + .5) / 1000
LOCATE 8, 16: PRINT "Terminal Speed           :"; INT((y0 / COS(ATN(z0))) * 100 + .5) / 100; " m/s"
LOCATE 9, 16: PRINT "Terminal Angle           :"; aT; " degree"
LOCATE 10, 16: PRINT "Coordinates of Vertex   :"; "("; INT((xm) * 10 + .5) / 10; ","; INT((ym) * 10 + .5) / 10; ")"
LOCATE 11, 16: PRINT "Cross-Wind Deflection   :"; INT((cw * (w0 - xx / (voo * COS(A * 3.14159265# / 180)))) * 100 + .5) / 100
LOCATE 13, 16: PRINT "Location of TARGET      :"; "("; ((xT) * 10 + 0.5) / 10; ((yT - vo) * 100 + 0.5) / 100; ")"
LOCATE 14, 16: PRINT "Location of FIREARM     :"; "(0,"; vo; ")"
LOCATE 15, 16: PRINT "Distance to TARGET      :"; (((xT ^ 2 + (yT - 2 * vo) ^ 2) ^ .5) * 10 + 0.5) / 10
COLOR 7
LOCATE 24, 11: PRINT " Pres [P] to repeat [Esc] to end ";
cc$ = INPUT$(1)
IF cc$ = CHR$(27) THEN SCREEN 9: CLS: END
END SUB


SUB InfHyres (xx, voo, vo1, ta, ta1, pa, pa1, ea, m, dm, tp, wind, xc, yc, yc1, vo, cw, xx1, koef, atm, G, GA, GS, TE, Pr, E, F, D, Tc, Incl, El)
LOCATE 4, 12: INPUT "Atmosphere: ICAO = 2"; atm
CLS
IF atm = 2 THEN GOTO 200:
```

```
200 LOCATE 4, 2: INPUT "G-Function; WM = 1, SI = 2, M118LR = 5, M118BB =
6, 300Gr.= 7, BM2 = 8,"; GA
TE = 288.15: Pr = 760: Tc = 21.11
IF GA = 1 THEN E = 0.182304: F = 44.271043: D = 0.0000649481#
IF GA = 2 THEN E = 0.179117: F = 46.77305: D = 0.0000673536#
IF GA = 5 THEN E = 0.178659: F = 46.77305: D = 0.0000481097#
IF GA = 6 THEN E = 0.181256: F = 44.59324: D = 0.0000639859#
IF GA = 7 THEN E = 0.149642: F = 34.00946: D = 0.0000707213#
IF GA = 8 THEN E = 0.168240: F = 35.7491: D = 0.00012027#
CLS
GOTO 400:

400
LOCATE 3, 12: INPUT "Distance: Inclined = 1, Declined =2, Horizontal = 3 "; Incl
IF Incl = 1 THEN
    LOCATE 4, 16: INPUT "Inclined Distance [m]          = "; xx
    LOCATE 5, 16: INPUT "Elevation Angle                = "; El
END IF
IF Incl = 2 THEN
    LOCATE 4, 15: INPUT "Declined Distance [m]          = "; xx
    LOCATE 5, 16: INPUT "Depression Angle               = "; El
    xx = xx * COS(El * 3.141592654# / 180)
    El = 0
END IF
IF Incl = 3 THEN
    LOCATE 4, 15: INPUT "Horizontal Distance [m]        = "; xx
    El = 0
END IF
CLS

LOCATE 5, 13: INPUT "y-Coordinate of Muzzle [m]    = "; vo
yc = xx * SIN(El * 3.141592654# / 180)
yc1 = yc
xx = xx * COS(El * 3.141592654# / 180)
xc = xx
LOCATE 6, 13: INPUT "Departure Velocity [m/s]      = "; voo
LOCATE 7, 13: INPUT "Temperature of Air [C]        = "; ta
LOCATE 8, 13: INPUT "Propellant Temperature [C]    = "; tp
LOCATE 9, 13: INPUT "Atmospheric Pressure [mm Hg]  = "; pa
LOCATE 10, 13: INPUT "Hunidity of Air [Decimal]    = "; ea
LOCATE 11, 13: INPUT "Projectile Standard Mass [kg] = "; m
LOCATE 12, 13: INPUT "Change in Projectile Mass    = "; dm
LOCATE 13, 13: INPUT "Range-Wind [m/s]             = "; wind
LOCATE 14, 13: INPUT "Cross-Wind [m/s]             = "; cw
LOCATE 15, 13: INPUT "Ballistics Coefficient BC    = "; koef
CLS
```

```
ta = ta + 273.15
IF ta > 273.16 AND ta <= 327.15 THEN
    ea = ea * 7.50187 * EXP(19.04 * (1 - 280.07 / ta))
END IF
IF ta > 255.15 AND ta < 273.15 THEN
    ea = ea * 7.50187 * EXP(22.024 * (1 - 279.24 / ta))
END IF

pa1 = ta / (1 - .3785 * ea / pa)
vo1 = (voo - .4 * voo * (dm / m) + .001 * voo * (tp - Tc))
koef = koef * (1 - dm / m)
END SUB

SUB menu (cog, cof, xf, yf, xfu, yfu, t$)
COLOR cog, cof
LOCATE xf - 1, yf: PRINT t$
LOCATE xf, yf: PRINT "É" + STRING$(yfu - yf, 205) + "»";
FOR i = xf + 1 TO xfu
    LOCATE i, yf: PRINT "º" + SPACE$(yfu - yf) + "º";
NEXT
LOCATE xfu + 1, yf: PRINT "È" + STRING$(yfu - yf, 205) + "¼";
END SUB

SUB NPkoef (k, l, r, q, h, y1, z1, v1, w1)
k = h * y1: l = h * z1
r = h * v1: q = h * w1
END SUB

SUB NPxyzvw (nk, x, x0, y, y0, z, z0, v, v0, w, w0, h, h0, k, l, r, q)
IF nk = 1 THEN
    x = x0: y = y0: z = z0
    v = v0: w = w0: h = h0
    GOTO fund:
END IF

IF nk = 2 OR nk = 3 THEN
    x = x0 + (.5 * h): y = y0 + (.5 * k)
    z = z0 + (.5 * l): v = v0 + (.5 * r)
    w = w0 + (.5 * q)
    GOTO fund:
END IF

IF nk = 4 THEN
    x = x0 + h: y = y0 + k: z = z0 + l
    v = v0 + r: w = w0 + q
END IF
```

fund:
END SUB

SUB y0z0 (y0, z0, A, vo1, wind)
y0 = SQR(vo1 ^ 2 + wind ^ 2 - 2 * vo1 * wind * COS(A * 3.141592654# / 180))
y0 = y0 * COS(A * 3.141592654# / 180)
z0 = TAN(A * 3.141592654# / 180)
z0 = z0 / (1 - wind / (vo1 * COS(A * 3.141592654# / 180)))
END SUB

SUB y1z1v1w1 (x, y, z, v, w, y1, z1, v1, w1, koef, pa1, wind, ys, yy, pa, ta1, TE, Pr, E, F, D)

ta1 = (TE / pa1) ^ .5
yy = y * SQR(1 + z ^ 2)
IF yy > 256! THEN
    y1 = -1 * koef * (pa / Pr) * ta1 * ((pa1 - .006328 * v) / pa1) ^ 4.4 * (E * yy - F) / yy
ELSE
    y1 = -1 * koef * (pa / Pr) * ta1 * ((pa1 - .006328 * v) / pa1) ^ 4.4 * D * (yy) ^ 2 / yy
END IF
z1 = -9.80665 / y ^ 2
v1 = z
w1 = 1 / y

END SUB

APPENDIX E

QBasic PC Program
RPROJ16.BAS
Non Standard Atmosphere
Reference G-functions G1, G2, ... G7, etc

'Given: Distance to Target, Elevation angle, super elevation angle, initial velocity, Ballistics Coefficient
'   Temperature of air, pressure and humidity are known;
'   Projectile flight is in presence of wind.(range-wind and cross-wind are known)
'Find: All the elements of the trajectory at an expected impact Point
'   Projectile Drop when departure angle is zero
'--
'CONTROL DATA
'
'NOTE The Input data belong to GB852 Lapua Scenar Bullet
'   (velocity 830 m/s, BC7 = 3.772 related to reference G7-Function)

'EXAMPLE 1 Reference G7-function. Horizontal shooting
'   Select ICAO atmosphere
'   Select G7 = 7
'   Select: Horizontal Shooting (Press 3)
'
'INPUT: Range = 1500, Departure angle = 1.1685, projectile velocity = 830
'   Temperature of air = 15,Temperature of propellant= 21.11
'   Presure = 760, Humidity of Air = 0.00, projectile mass = 1, change in mass = 0
'   Range wind = 0.0, cross wind, 0.00, BC = 3.772

'RESULTS: Coordinats of Impact Point (1500, 0.33), Distance to target = 1500
'   Vertical Deviation from the center of Target = 0.33m, Time of Flight = 2.84s
'   Impact Speed = 348.88 m/s, Impact Angle = - 2.051 Degree
'   Coordinates of the trajectory vertex (869, 10.3), wind deflection 0 m

' EXAMPLE 2
' Select ICAO atmosphere
' Select G7 = 7
' Select: Inclined Shooting (Press 1)

'INPUT: Inclined Distance = 1500, Elevation Angle = 30, Super Elevation Angle = 0.8552
'   y- coordinate of Muzzle = 0, Pprojectile velocity = 830
'   Temperature of air = 15,Temperature of propellant= 21.11
'   Presure = 760, Humidity of Air = 0.00, projectile mass = 1, change in mass = 0
'   Range wind = 4.5, cross wind= 0, BC = 3.772
'
'RESULTS: Coordinates of Point of Impact (1298.97, 750.1), Distance to target = 1500.19
'   Vertical Deviation from center of Target = 0.56m, Time of Flight = 2.807s
'   Impact Speed = 351.35 m/s, Impact Angle = 28.259 Degree
'   Coordinates of the trajectory vertex (0, 0): No Vertex, Cross-wind deflection 0.00 m
'
'--

'Functions & Subs.

DECLARE SUB y1z1v1w1 (x, y, z, v, w, y1, z1, v1, w1, koef, pa1, wind, ys, yy, pa, ta1, TE, E, F, D, Pr)
DECLARE SUB InfHyres (x0, y0, z0, v0, w0, a, h0, ta, pa, ea, tp, ta1, pa1, xx1, voo, vo1, wind, koef, cw, vv, E, F, D, Pr, TE, m, dm, xax, Incl, El)
DECLARE SUB NPxyzvw (nk, x, x0, y, y0, z, z0, v, v0, w, w0, h, h0, k, L, r, q)
DECLARE SUB NPkoef (k, L, r, q, h, y1, z1, v1, w1)
DECLARE SUB menu (cog, cof, xf, yf, xfu, yfu, t$)
DECLARE SUB c (koef, m, dm, BC)

'Variables
DIM m(4, 4), v(4)
rendi = 4
cog = 7: cof = 0

'Zgjidhja
CLS
fillimi:
menu cog, cof, 3, 10, 21, 70, "INITIAL DATA"

InfHyres x0, y0, z0, v0, w0, a, h0, ta, pa, ea, tp, ta1, pa1, xx1, voo, vo1, wind, koef, cw, vv, E, F, D, Pr, TE, m, dm, xax, Incl, El
c koef, m, dm, BC

F:
  FOR nk = 1 TO rendi
  NPxyzvw nk, x, x0, y, y0, z, z0, v, v0, w, w0, h, h0, k, L, r, q
  y1z1v1w1 x, y, z, v, w, y1, z1, v1, w1, koef, pa1, wind, ys, yy, pa, ta1, TE, E, F, D, Pr
  NPkoef k, L, r, q, h, y1, z1, v1, w1
  m(nk, 1) = k: m(nk, 2) = L
  m(nk, 3) = r: m(nk, 4) = q
NEXT nk

'Calculation
FOR i = 1 TO rendi
  v(i) = 1 / 6 * (m(1, i) + 2 * m(2, i) + 2 * m(3, i) + m(4, i))
NEXT i

'New Data
x0 = x0 + h: y0 = y0 + v(1): z0 = z0 + v(2)
v0 = v0 + v(3): w0 = w0 + v(4)

IF ABS(z0) < .00001 THEN
  ymax = v0 + wind * w0 * z0
  xmax = x0 + wind * w0
END IF
IF Incl = 1 THEN
  yyT = v0 + wind * w0 * z0
ELSE
  yyT = v0
END IF
  xxT = x0 + wind * w0
IF (xxT - xax) <= .001 THEN
  tt = w0
  xt = xxT
  yyT = yyT
  yt = v0
  at = 180 * ATN(z0) / 3.141592654#
  vt = y0 * (1 + z0 ^ 2) ^ .5
zt = cw * (w0 - xt / (voo * COS(a * 3.14159265# / 180)))
END IF

ytt = yt
IF x0 >= 10 AND ABS(v0 - ytt) >= .1 THEN

  'Display Results

menu cog, cof, 6, 20, 22, 72, "RESULTS:"

  LOCATE 10, 25: PRINT "Impact Point Coordinates   = "; "("; INT((xt) * 100 + .5) / 100; ","; INT((yyT - vv) * 100 + .5) / 100; ")"
  LOCATE 11, 25: PRINT "Location of Firearm   = "; (0); ","; (vv)
  LOCATE 12, 25: PRINT "Inclined Range   = "; INT(((xt ^ 2 + (2 * vv - yyT) ^ 2) ^ 0.5) * 100 + .5) / 100
  LOCATE 13, 25: PRINT "Vertical Deviation from Target = "; INT((yyT - vv - xt * TAN(El * 3.14159265 / 180)) * 1000 + .5) / 1000
  LOCATE 14, 25: PRINT "Time [s]   = "; INT((tt) * 1000 + .5) / 1000
  LOCATE 15, 25: PRINT "Corresponding Velocity   = "; INT((vt) * 100 + .5) / 100
  LOCATE 16, 25: PRINT "Corresponding Angle [Deg] = "; INT((at) * 1000 + .5) / 1000
  LOCATE 17, 25: PRINT "Cross-Wind Deflection   = "; INT((zt) * 100 + .5) / 100
  LOCATE 18, 25: PRINT "Trajectory Vertex [m]   = "; "("; INT((xmax) + .5); ","; INT((ymax) * 10 + .5) / 10; ")"
  LOCATE 19, 25: PRINT "Ballistics Coefficient   = "; BC
  LOCATE 24, 8: PRINT "Impact Coordinates: Ref. System Origin at Muzzle = "; "("; INT((xt) * 100 + .5) / 100; ","; INT((yyT - vv) * 100 + .5) / 100; ")"
ELSE
  GOTO F:
END IF
END

SUB c (koef, m, dm, BC)
BC = koef
koef = koef * (1 - dm / m)
END SUB

SUB InfHyres (x0, y0, z0, v0, w0, a, h0, ta, pa, ea, tp, ta1, pa1, xx1, voo, vo1, wind, koef, cw, vv, E, F, D, Pr, TE, m, dm, xax, Incl, El)
LOCATE 4, 15: INPUT "ATMOSPHERE: ASM = 1, ICAO = 2; TSA = 3"; atm
CLS
IF atm = 1 THEN GOTO 100:
IF atm = 2 THEN GOTO 200:
IF atm = 3 THEN GOTO 300:

100 LOCATE 4, 13: INPUT "G-Function: G1 = 1, G2 = 2, G5 = 5, G6 = 6, G7 = 7, G8 = 8 "; G
TE = 289.6: Pr = 750
IF G = 1 THEN E = .312914: F = 79.3976: D = .000100347#
IF G = 2 THEN E = .1413: F = 29.9097: D = .00009116233#
IF G = 5 THEN E = .19734: F = 49.0806: D = .0000743571#
IF G = 6 THEN E = .140533: F = 23.6633: D = .00010334#
IF G = 7 THEN E = .150355: F = 34.7319: D = .000056648#
IF G = 8 THEN E = .14733: F = 29.085: D = .000099366#

CLS
GOTO 400:

200 LOCATE 4, 13: INPUT "G-Function: G1 = 1, G2 = 2, G5 = 5, G6 = 6, G7 = 7, G8 = 8 "; GA
TE = 288.15: Pr = 760
IF GA = 1 THEN E = .31574: F = 78.6769: D = .00010584#
IF GA = 2 THEN E = .143353: F = 30.2415: D = .00009287#
IF GA = 5 THEN E = .200207: F = 49.625: D = .000075244#
IF GA = 6 THEN E = .142352: F = 23.6937: D = .000105244#
IF GA = 7 THEN E = .152593: F = 35.1717: D = .000057679#
IF GA = 8 THEN E = .149441: F = 29.379: D = .000101154#
IF GA = 10 THEN E = .325383: F = 83.6082: D = .00010724#
CLS
GOTO 400:

300 LOCATE 4, 13: INPUT "G-Function: Siacci = 1, Russian G43-Function = 3"; GS
TE = 289.08
Pr = 750
IF GS = 1 THEN E = .3333333#: F = 80: D = .000212
IF GS = 2 THEN E = .320243: F = 81.3721: D = .00010807#
IF GS = 3 THEN E = .157713: F = 36.39542: D = .00007454#
CLS
GOTO 400:

400
LOCATE 3, 12: INPUT "Distance: Inclined = 1, Declined = 2, Horizontal = 3 "; Incl
IF Incl = 1 THEN
   LOCATE 4, 16: INPUT "Inclined Distance [m]   = "; xax
   LOCATE 5, 16: INPUT "Elevation Angle   = "; El
   xax = xax * COS(El * 3.141592654# / 180)
   LOCATE 6, 16: INPUT "Super Elevation Angle   = "; z0
END IF
IF Incl = 2 THEN
   LOCATE 4, 15: INPUT "Declined Range [m]   = "; xax
   LOCATE 5, 16: INPUT "Depression Angle   = "; El
   xax = xax * COS(El * 3.141592654# / 180)
   LOCATE 6, 15: INPUT "Departure angle   = "; z0
   El = 0
END IF
IF Incl = 3 THEN
   LOCATE 4, 15: INPUT "Horizontal Range [m]   = "; xax
   LOCATE 5, 15: INPUT "Departure angle   = "; z0
   El = 0
END IF
CLS

```
LOCATE 4, 13: INPUT "y-coordinate of Muzzle     = "; v0
z0 = z0 + El
LOCATE 5, 13: INPUT "Departure Velocity [m/s]   = "; y0
LOCATE 6, 13: INPUT "Temperature of Air [C]     = "; ta
LOCATE 7, 13: INPUT "Propellant Temperature[C]  = "; tp
LOCATE 8, 13: INPUT "Atmospheric Pressure [mm]  = "; pa
LOCATE 9, 13: INPUT "Hunidity of Air [Decimal]  = "; ea
LOCATE 10, 13: INPUT "Projectile Mass           = "; m
LOCATE 11, 13: INPUT "Change in Projectile mass = "; dm
LOCATE 12, 13: INPUT "Range Wind                = "; wind
LOCATE 13, 13: INPUT "Cross Wind                = "; cw
LOCATE 14, 13: INPUT "Ballistics Coefficient    = "; koef
LOCATE 15, 10: INPUT "Integration Step 10, 1, 0.1 = "; h0
vv = v0: a = z0: voo = y0
ta = ta + 273.15
IF ta > 273.16 AND ta <= 327.15 THEN
   ea = ea * 7.50187 * EXP(19.04 * (1 - 280.07 / ta))
END IF
IF ta > 255.15 AND ta < 273.15 THEN
   ea = ea * 7.50187 * EXP(22.024 * (1 - 279.24 / ta))
END IF
pa1 = ta / (1 - .3785 * ea / pa)
vo1 = (voo - .4 * voo * (dm / m) + .001 * voo * (tp - 15))
y0 = SQR(vo1 ^ 2 + wind ^ 2 - 2 * vo1 * wind * COS(a * 3.141592654# / 180))
y0 = y0 * COS(a * 3.141592654# / 180)
z0 = TAN(a * 3.141592654# / 180)
z0 = z0 / (1 - wind / (vo1 * COS(a * 3.141592654# / 180)))
CLS
END SUB

SUB menu (cog, cof, xf, yf, xfu, yfu, t$)

COLOR cog, cof
LOCATE xf - 1, yf: PRINT t$

LOCATE xf, yf: PRINT "É" + STRING$(yfu - yf, 205) + "»";

FOR i = xf + 1 TO xfu
   LOCATE i, yf: PRINT "º" + SPACE$(yfu - yf) + "º";
NEXT
LOCATE xfu + 1, yf: PRINT "È" + STRING$(yfu - yf, 205) + "¼";
END SUB

SUB NPkoef (k, L, r, q, h, y1, z1, v1, w1)

k = h * y1: L = h * z1
```

```
r = h * v1: q = h * w1
END SUB


SUB NPxyzvw (nk, x, x0, y, y0, z, z0, v, v0, w, w0, h, h0, k, L, r, q)

IF nk = 1 THEN
   x = x0: y = y0: z = z0
   v = v0: w = w0: h = h0
   GOTO fund:
END IF

IF nk = 2 OR nk = 3 THEN
   x = x0 + (.5 * h): y = y0 + (.5 * k)
   z = z0 + (.5 * L): v = v0 + (.5 * r)
   w = w0 + (.5 * q)
   GOTO fund:
END IF

IF nk = 4 THEN
   x = x0 + h: y = y0 + k: z = z0 + L
   v = v0 + r: w = w0 + q
END IF

fund:
END SUB


SUB y1z1v1w1 (x, y, z, v, w, y1, z1, v1, w1, koef, pa1, wind, ys, yy, pa, ta1, TE, E, F, D, Pr)

ta1 = (TE / pa1) ^ .5
yy = y * SQR(1 + z ^ 2)
IF yy * ta1 > 256! THEN
    y1 = -1 * koef * (pa / Pr) * ta1 * ((pa1 - .006328 * v) / pa1) ^ 4.4 * (E * yy - F) / yy
ELSE
    y1 = -1 * koef * ta1 * (pa / Pr) * ((pa1 - .006328 * v) / pa1) ^ 4.4 * D * (yy) ^ 2 / yy
END IF
z1 = -9.80665 / y ^ 2
v1 = z
w1 = 1 / y

END SUB
```

QBasic PC Program
RLAPUA16.BAS
Lapua 0.338 GB528 Scenar Bullet
Non Standard Atmosphere

'FIND : Elements of Trajectory at a given distance, as well as the projectile
' drop when departure angle is zero.
'GIVEN: Elevation Angle, Super elevation angle, Initial speed, BC

'Lapua Bullet GB528: Initial velocity = 830m/s, BC = 3.796

'---
'

'Find: Range (Point of impact), Time of Flight, Impact Speed, Impact Angle, etc.

'Given: Distance to Target, Elevation angle, super elevation angle, initial velocity, Ballistics Coefficient
' Temperature of air, pressure and humidity are known;
' Projectile flight is in presence of wind.(range-wind and cross-wind are known)
'Find: All the elements of the trajectory at an expected impact Point
' Projectile Drop when departure angle is zero

'Control Data
' Example 1. Inclined Shooting
'Select: ICAO atmosphere: 2
'Select Characteristic G-function GB528 Lapua

'Input: y-coordinate of FIREARM = 0, Elevation angle 30 degree; Super Elevation Angle = 0.9934,
' Inclined Range = 1500m, Departure speed = 830
' Temperature of Air, 15, temperature of propellant, 21.11;
' Pressure = 760, humodity:0.00, Projectile Mass, 1;
' Change in Projectile mass = 0, range wind, 4.5; cross wind, 0; BC = 3.796
'

'Results: X-coordinate = 1299, y-coordinate = 750.091, inclined Range = 1500.05m;
' vertical deviation from center of target = 0.11m, time = 2.792s, Impact velocity = 352.3m/s

'Example 2 Departure angle = 0. Projectile Drop
' Select Horizontal Shooting and repeat all other steps. Departure angle = 0.
'Results: Coordinates of Impact at 1500 meter: (1499.98, -30.00m), Distance to the impact point = 1500.28
' Vertical deviation from center of target (Projectile Drop)= - 30.005, Time 2.825, velocity 349.21 m/s

'Note: When elevation angle is zero, the inclined range is the horizontal range.
'---

'Functions & Subs.

'Functions & Subs.

DECLARE SUB y1z1v1w1 (x, y, z, v, w, y1, z1, v1, w1, koef, pa1, wind, ys, yy, pa, ta1, TE, E, F, D, D1, Pr)
DECLARE SUB InfHyres (x0, y0, z0, v0, w0, a, h0, ta, pa, ea, tp, ta1, pa1, xx1, voo, vo1, wind, koef, cw, vv, E, F, D, Pr, TE, m, dm, xax, Incl, El)
DECLARE SUB NPxyzvw (nk, x, x0, y, y0, z, z0, v, v0, w, w0, h, h0, k, L, r, q)
DECLARE SUB NPkoef (k, L, r, q, h, y1, z1, v1, w1)
DECLARE SUB menu (cog, cof, xf, yf, xfu, yfu, t$)
DECLARE SUB c (koef, m, dm, BC)

'Variables

DIM m(4, 4), v(4)
rendi = 4
cog = 7: cof = 0

'Zgjidhja
CLS

fillimi:

menu cog, cof, 3, 10, 21, 70, "INITIAL DATA"
InfHyres x0, y0, z0, v0, w0, a, h0, ta, pa, ea, tp, ta1, pa1, xx1, voo, vo1, wind, koef, cw, vv, E, F, D, Pr, TE, m, dm, xax, Incl, El
c koef, m, dm, BC

F:
FOR nk = 1 TO rendi
    NPxyzvw nk, x, x0, y, y0, z, z0, v, v0, w, w0, h, h0, k, L, r, q
    y1z1v1w1 x, y, z, v, w, y1, z1, v1, w1, koef, pa1, wind, ys, yy, pa, ta1, TE, E, F, D, D1, Pr
    NPkoef k, L, r, q, h, y1, z1, v1, w1
    m(nk, 1) = k: m(nk, 2) = L
    m(nk, 3) = r: m(nk, 4) = q
NEXT nk

'Calculation

FOR i = 1 TO rendi
    v(i) = 1 / 6 * (m(1, i) + 2 * m(2, i) + 2 * m(3, i) + m(4, i))
NEXT i

'New Data

x0 = x0 + h: y0 = y0 + v(1): z0 = z0 + v(2)

```
v0 = v0 + v(3): w0 = w0 + v(4)

IF ABS(z0) < .00001 THEN
   ymax = v0 + wind * w0 * z0
   xmax = x0 + wind * w0
END IF
IF Incl = 1 THEN
   yyT = v0 + wind * w0 * z0
ELSE
   yyT = v0
END IF
xxT = x0 + wind * w0
IF (xxT - xax) <= .001 THEN
   tt = w0
   xt = xxT
   yyT = yyT
   yt = v0
   at = 180 * ATN(z0) / 3.141592654#
   vt = y0 * (1 + z0 ^ 2) ^ .5
zt = cw * (w0 - xt / (voo * COS(a * 3.14159265# / 180)))
END IF


ytt = yt
IF x0 >= 10 AND ABS(v0 - ytt) >= .1 THEN



'Display Results

   menu cog, cof, 6, 20, 22, 72, "RESULTS:"
      LOCATE 10, 25: PRINT "Impact Point Coordinates   = "; "("; INT((xt) * 100 + .5)
/ 100; ","; INT((yyT - vv) * 100 + .5) / 100; ")"
      LOCATE 11, 25: PRINT "Location of Firearm        = "; (0); ","; (vv)
      LOCATE 12, 25: PRINT "Inclined Range             = "; INT(((xt ^ 2 + (2 * vv - yyT)
^ 2) ^ 0.5) * 100 + .5) / 100
      LOCATE 13, 25: PRINT "Vertical Deviation from Target = "; INT((yyT - vv - xt *
TAN(El * 3.14159265 / 180)) * 1000 + .5) / 1000
      LOCATE 14, 25: PRINT "Time [s]                   = "; INT((tt) * 1000 + .5) / 1000
      LOCATE 15, 25: PRINT "Corresponding Velocity     = "; INT((vt) * 100 + .5) / 100
      LOCATE 16, 25: PRINT "Corresponding Angle [Deg]  = "; INT((at) * 1000 + .5)
/ 1000
      LOCATE 17, 25: PRINT "Cross-Wind Deflection      = "; INT((zt) * 100 + .5) / 100
      LOCATE 18, 25: PRINT "Trajectory Vertex [m]      = "; "("; INT((xmax) + .5); ","; 
INT((ymax) * 10 + .5) / 10; ")"
      LOCATE 19, 25: PRINT "Ballistics Coefficient     = "; BC
      LOCATE 24, 8: PRINT "Impact Coordinates: Ref. System Origin at Muzzle = ";
"("; INT((xt) * 100 + .5) / 100; ","; INT((yyT - vv) * 100 + .5) / 100; ")"
```

ELEMENTS OF EXTERIOR BALLISTICS | 255

```
  ELSE
    GOTO F:
  END IF
END

SUB c (koef, m, dm, BC)
BC = koef
koef = koef * (1 - dm / m)

END SUB

SUB InfHyres (x0, y0, z0, v0, w0, a, h0, ta, pa, ea, tp, ta1, pa1, xx1, voo, vo1, wind, koef,
cw, vv, E, F, D, Pr, TE, m, dm, xax, Incl, El)

LOCATE 4, 16: INPUT "Atmosphere: ASM = 1, ICAO = 2 "; atm
CLS
IF atm = 1 THEN GOTO 100:
IF atm = 2 THEN GOTO 200:

100 LOCATE 4, 5: INPUT "Lapua G-Function GB528 = 1,"; G
TE = 289.6: Pr = 750: Tc = 21.11
IF G = 1 THEN E = 0.13895: F = 29.709: D = 0.0242: D1 = 1.399#
CLS
GOTO 400:

200 LOCATE 4, 5: INPUT "G-Function: Lapua GB528 = 1,"; GA
TE = 288.15: Pr = 760: Tc = 21.11

IF GA = 1 THEN E = 0.141: F = 30.031: D = 0.02438: D1 = 1.3696#
CLS
GOTO 400:

400
LOCATE 3, 12: INPUT "Distance: Inclined = 1, Declined = 2, Horizontal = 3 "; Incl
IF Incl = 1 THEN
    LOCATE 4, 16: INPUT "Inclined Distance [m]       = "; xax
    LOCATE 5, 16: INPUT "Elevation Angle             = "; El
    xax = xax * COS(El * 3.141592654# / 180)
    LOCATE 6, 16: INPUT "Super Elevation Angle       = "; z0
END IF
IF Incl = 2 THEN
    LOCATE 4, 15: INPUT "Declined Range [m]          = "; xax
    LOCATE 5, 16: INPUT "Depression Angle            = "; El
    xax = xax * COS(El * 3.141592654# / 180)
    LOCATE 6, 15: INPUT "Departure angle             = "; z0
    El = 0
```

```
END IF
IF Incl = 3 THEN
    LOCATE 4, 15: INPUT "Horizontal Range [m]        = "; xax
    LOCATE 5, 15: INPUT "Departure angle             = "; z0
    El = 0
END IF
CLS
LOCATE 4, 13: INPUT "y-coordinate of Muzzle    = "; v0
z0 = z0 + El
LOCATE 5, 13: INPUT "Departure Velocity [m/s]  = "; y0
LOCATE 6, 13: INPUT "Temperature of Air [C]    = "; ta
LOCATE 7, 13: INPUT "Propellant Temperature[C] = "; tp
LOCATE 8, 13: INPUT "Atmospheric Pressure [mm] = "; pa
LOCATE 9, 13: INPUT "Hunidity of Air [Decimal] = "; ea
LOCATE 10, 13: INPUT "Projectile Mass          = "; m
LOCATE 11, 13: INPUT "Change in Projectile mass = "; dm
LOCATE 12, 13: INPUT "Range Wind               = "; wind
LOCATE 13, 13: INPUT "Cross Wind               = "; cw
LOCATE 14, 13: INPUT "Ballistics Coefficient   = "; koef
LOCATE 15, 10: INPUT "Integration Step 10, 1, 0.1 = "; h0
vv = v0: a = z0: voo = y0
ta = ta + 273.15
IF ta > 273.16 AND ta <= 327.15 THEN
    ea = ea * 7.50187 * EXP(19.04 * (1 - 280.07 / ta))
END IF
IF ta > 255.15 AND ta < 273.15 THEN
    ea = ea * 7.50187 * EXP(22.024 * (1 - 279.24 / ta))
END IF
pa1 = ta / (1 - .3785 * ea / pa)
vo1 = (voo - .4 * voo * (dm / m) + .001 * voo * (tp - Tc))
y0 = SQR(vo1 ^ 2 + wind ^ 2 - 2 * vo1 * wind * COS(a * 3.141592654# / 180))
y0 = y0 * COS(a * 3.141592654# / 180)
z0 = TAN(a * 3.141592654# / 180)
z0 = z0 / (1 - wind / (vo1 * COS(a * 3.141592654# / 180)))
CLS
END SUB

SUB menu (cog, cof, xf, yf, xfu, yfu, t$)

COLOR cog, cof
LOCATE xf - 1, yf: PRINT t$

LOCATE xf, yf: PRINT "É" + STRING$(yfu - yf, 205) + "»";

FOR i = xf + 1 TO xfu
    LOCATE i, yf: PRINT "º" + SPACE$(yfu - yf) + "º";
```

NEXT
LOCATE xfu + 1, yf: PRINT "È" + STRING$(yfu - yf, 205) + "¼";
END SUB

SUB NPkoef (k, L, r, q, h, y1, z1, v1, w1)

k = h * y1: L = h * z1
r = h * v1: q = h * w1
END SUB

SUB NPxyzvw (nk, x, x0, y, y0, z, z0, v, v0, w, w0, h, h0, k, L, r, q)

IF nk = 1 THEN
   x = x0: y = y0: z = z0
   v = v0: w = w0: h = h0
   GOTO fund:
END IF

IF nk = 2 OR nk = 3 THEN
   x = x0 + (.5 * h): y = y0 + (.5 * k)
   z = z0 + (.5 * L): v = v0 + (.5 * r)
   w = w0 + (.5 * q)
   GOTO fund:
END IF

IF nk = 4 THEN
   x = x0 + h: y = y0 + k: z = z0 + L
   v = v0 + r: w = w0 + q
END IF

fund:
END SUB

SUB y1z1v1w1 (x, y, z, v, w, y1, z1, v1, w1, koef, pa1, wind, ys, yy, pa, ta1, TE, E, F, D, D1, Pr)

ta1 = (TE / pa1) ^ .5
yy = y * SQR(1 + z ^ 2)
IF yy * ta1 > 325! THEN
   y1 = -1 * koef * (pa / Pr) * ta1 * ((pa1 - .006328 * v) / pa1) ^ 4.4 * (E * yy - F) / yy
ELSE
   y1 = -1 * koef * (pa / Pr) * ta1 * ((pa1 - .006328 * v) / pa1) ^ 4.4 * (D * yy - D1) / yy
END IF
z1 = -9.80665 / y ^ 2
v1 = z
w1 = 1 / y

END SUB

ately # QBasic PC Program
RCHA16.BAS
Non Standard Atmosphere

'Given: Distance to Target, Elevation angle, super elevation angle, initial velocity, Ballistics Coefficient
' Temperature of air, pressure and humidity are known;
' Projectile flight is in presence of wind.(range-wind and cross-wind are known)
'Find: The point of impact and all the elements of the trajectory at the he given distance to target
' Projectile Drop when departure angle is zero
'---
"
' Ballistics Coefficient is Known
' Non Standard Atmosphere, Wind Present

' LIST OF BULLETS
' ICAO Atmosphere Characteristic G-functions
' 1. Bullet 0.300 Winchester Magnum, Velocity 884m/s, BC = 4.97096: WM = 1 (Press 1)
' 2. Caliber 0.308, 168 grain, Sierra International Bullet, Velocity 792.48m/s, BC = 5.6325: SI = 2 (press 2)
' 3. M118LR Bullet, 0.308" mass 0.01134 kg. Velocity 884m/s, BC = 5.3970 M118LR = 5 (Press 5)
' 4. M118 Ball Bullet (Federal GM308M2), 0.308", mass = 0.00134kg, velocity 792.48m/s, BC =5.3973 'M118BB = 6 (Press 6)
' 5. 300 gr. .338 - .416 Bullet (mass 0.01944kg, Velocity 927.40m/s, BC = 3.791 300Gr = 7 (Press 7)
' 6. caliber 0.30 Ball M2 Bullet, mass 0.00972 kg, Velocity 853.44 m/s, BC =6.2965 BM2 = 8 (Press 8)

' Select: ICAO atmosphere: 2
' Select Characteristic G-function M118LR : 5

' Example 1 Uphill Shooting
' Input: Inclined = 1, Inclined range = 1500, Elevation angle= 30 degree; Super elevation Angle = 1.2986,
'   y-coordinate of muzzle = 0, Departure speed = 884
'   Temperature of Air, 15, temperature of propellant, 21.11,
'   Pressure = 750, humodity = 0.78, Projectile Mass, 1,
'   Change in Projectile mass = 0, range wind, 0; cross wind, 0; BC = 5.397
'
' Results: Coordinats of point of impact (1299.03, 750.36), Inclined range 1500.17
'   Vertical deviation from center of target = 0.362m, time = 3.33s, Impact velocity = 287m/s
'   Impact Angle = 27.28 degree.

' Example 2 Use the data in example1 to find the drop of projectile at 1000m and 1500m if the projectile is lunched
'   at an angle = 0 degree.
'   FOLLOW The same steps as above only the departure angle must be equal to zero degree

' Results: x-coordinate = 1000, y-coordinate = -12.01; time = 1.779: Impact velocity = 376.88
'   X-coordinate = 1500, y-coordinate = -39.4; time = 3.334: Impact velocity = 288.5

'Functions & Subs.
DECLARE SUB y1z1v1w1 (x, y, z, v, w, y1, z1, v1, w1, koef, pa1, wind, ys, yy, pa, ta1, TE, E, F, D, Pr)
DECLARE SUB InfHyres (x0, y0, z0, v0, w0, a, h0, ta, pa, ea, tp, ta1, pa1, xx1, voo, vo1, wind, koef, cw, vv, E, F, D, Pr, TE, m, dm, xax, Incl, El)
DECLARE SUB NPxyzvw (nk, x, x0, y, y0, z, z0, v, v0, w, w0, h, h0, k, L, r, q)
DECLARE SUB NPkoef (k, L, r, q, h, y1, z1, v1, w1)
DECLARE SUB menu (cog, cof, xf, yf, xfu, yfu, t$)
DECLARE SUB c (koef, m, dm, BC)

'Variables
DIM m(4, 4), v(4)
rendi = 4
cog = 7: cof = 0
'Zgjidhja
CLS
fillimi:
menu cog, cof, 3, 10, 21, 70, "INITIAL DATA"
InfHyres x0, y0, z0, v0, w0, a, h0, ta, pa, ea, tp, ta1, pa1, xx1, voo, vo1, wind, koef, cw, vv, E, F, D, Pr, TE, m, dm, xax, Incl, El
c koef, m, dm, BC

F:
FOR nk = 1 TO rendi
    NPxyzvw nk, x, x0, y, y0, z, z0, v, v0, w, w0, h, h0, k, L, r, q
    y1z1v1w1 x, y, z, v, w, y1, z1, v1, w1, koef, pa1, wind, ys, yy, pa, ta1, TE, E, F, D, Pr
    NPkoef k, L, r, q, h, y1, z1, v1, w1
    m(nk, 1) = k: m(nk, 2) = L
    m(nk, 3) = r: m(nk, 4) = q
NEXT nk
'Calculation
FOR i = 1 TO rendi
v(i) = 1 / 6 * (m(1, i) + 2 * m(2, i) + 2 * m(3, i) + m(4, i))
NEXT i

'New Data
x0 = x0 + h: y0 = y0 + v(1): z0 = z0 + v(2)
v0 = v0 + v(3): w0 = w0 + v(4)
IF ABS(z0) < .00001 THEN
   ymax = v0 + wind * w0 * z0
   xmax = x0 + wind * w0
END IF
IF Incl = 1 THEN
   yyT = v0 + wind * w0 * z0
ELSE
   yyT = v0
END IF
xxT = x0 + wind * w0
IF (xxT - xax) <= .001 THEN
   tt = w0
   xt = xxT
   yyT = yyT
   yt = v0
   at = 180 * ATN(z0) / 3.141592654#
   vt = y0 * (1 + z0 ^ 2) ^ .5
zt = cw * (w0 - xt / (voo * COS(a * 3.14159265# / 180)))
END IF

ytt = yt
IF x0 >= 10 AND ABS(v0 - ytt) >= .1 THEN
   'Display Results
   menu cog, cof, 6, 20, 22, 72, "RESULTS:"
   LOCATE 10, 25: PRINT "Impact Point Coordinates   = "; "("; INT((xt) * 100 + .5) / 100; ","; INT((yyT - vv) * 100 + .5) / 100; ")"
   LOCATE 11, 25: PRINT "Location of Firearm   = "; (0); ","; (vv)
   LOCATE 12, 25: PRINT "Inclined Range   = "; INT(((xt ^ 2 + (2 * vv - yyT) ^ 2) ^ 0.5) * 100 + .5) / 100
   LOCATE 13, 25: PRINT "Vertical Deviation from Target = "; INT((yyT - vv - xt * TAN(El * 3.14159265 / 180)) * 1000 + .5) / 1000
   LOCATE 14, 25: PRINT "Time [s]   = "; INT((tt) * 1000 + .5) / 1000
   LOCATE 15, 25: PRINT "Corresponding Velocity   = "; INT((vt) * 100 + .5) / 100
   LOCATE 16, 25: PRINT "Corresponding Angle [Deg] = "; INT((at) * 1000 + .5) / 1000
   LOCATE 17, 25: PRINT "Cross-Wind Deflection   = "; INT((zt) * 100 + .5) / 100
   LOCATE 18, 25: PRINT "Trajectory Vertex [m]   = "; "("; INT((xmax) + .5); ","; INT((ymax) * 10 + .5) / 10; ")"
   LOCATE 19, 25: PRINT "Ballistics Coefficient   = "; BC
   LOCATE 24, 8: PRINT "Impact Coordinates: Ref. System Origin at Muzzle = "; "("; INT((xt) * 100 + .5) / 100; ","; INT((yyT - vv) * 100 + .5) / 100; ")"
ELSE
   GOTO F:

```
END IF
END
SUB c (koef, m, dm, BC)
BC = koef
koef = koef * (1 - dm / m)
END SUB

SUB InfHyres (x0, y0, z0, v0, w0, a, h0, ta, pa, ea, tp, ta1, pa1, xx1, voo, vo1, wind, koef,
cw, vv, E, F, D, Pr, TE, m, dm, xax, Incl, El)
LOCATE 4, 16: INPUT "Atmosphere: ICAO = 2 "; atm
CLS

IF atm = 2 THEN GOTO 200:
200 LOCATE 4, 5: INPUT "G-Function; WM = 1, SI = 2, M118LR = 5, M118BB =
6, 300Gr.= 7, BM2 = 8,"; GA
TE = 288.15: Pr = 760: Tc = 21.11
IF GA = 1 THEN E = 0.182304: F = 44.271043: D = 0.0000649481#
IF GA = 2 THEN E = 0.179117: F = 46.77305: D = 0.0000673536#
IF GA = 5 THEN E = 0.178659: F = 46.77305: D = 0.0000481097#
IF GA = 6 THEN E = 0.181256: F = 44.59324: D = 0.0000639859#
IF GA = 7 THEN E = 0.149642: F = 34.00946: D = 0.0000707213#
IF GA = 8 THEN E = 0.168240: F = 35.7491: D = 0.00012027#
CLS
GOTO 400:

400
LOCATE 3, 12: INPUT "Distance: Inclined = 1, Declined = 2, Horizontal = 3 "; Incl
IF Incl = 1 THEN
    LOCATE 4, 16: INPUT "Inclined Distance [m]        = "; xax
    LOCATE 5, 16: INPUT "Elevation Angle              = "; El
    xax = xax * COS(El * 3.141592654# / 180)
    LOCATE 6, 16: INPUT "Super Elevation Angle        = "; z0
END IF
IF Incl = 2 THEN
    LOCATE 4, 15: INPUT "Declined Range [m]           = "; xax
    LOCATE 5, 16: INPUT "Depression Angle             = "; El
    xax = xax * COS(El * 3.141592654# / 180)
    LOCATE 6, 15: INPUT "Departure angle              = "; z0
    El = 0
END IF
IF Incl = 3 THEN
    LOCATE 4, 15: INPUT "Horizontal Range [m]         = "; xax
    LOCATE 5, 15: INPUT "Departure angle              = "; z0
    El = 0
END IF
CLS
```

```
LOCATE 4, 13: INPUT "y-coordinate of Muzzle      = "; v0
z0 = z0 + El
LOCATE 5, 13: INPUT "Departure Velocity [m/s]    = "; y0
LOCATE 6, 13: INPUT "Temperature of Air [C]      = "; ta
LOCATE 7, 13: INPUT "Propellant Temperature[C]   = "; tp
LOCATE 8, 13: INPUT "Atmospheric Pressure [mm]   = "; pa
LOCATE 9, 13: INPUT "Hunidity of Air [Decimal]   = "; ea
LOCATE 10, 13: INPUT "Projectile Mass            = "; m
LOCATE 11, 13: INPUT "Change in Projectile mass  = "; dm
LOCATE 12, 13: INPUT "Range Wind                 = "; wind
LOCATE 13, 13: INPUT "Cross Wind                 = "; cw
LOCATE 14, 13: INPUT "Ballistics Coefficient     = "; koef
LOCATE 15, 10: INPUT "Integration Step 10, 1, 0.1 = "; h0
vv = v0: a = z0: voo = y0
ta = ta + 273.15
IF ta > 273.16 AND ta <= 327.15 THEN
    ea = ea * 7.50187 * EXP(19.04 * (1 - 280.07 / ta))
END IF
IF ta > 255.15 AND ta < 273.15 THEN
    ea = ea * 7.50187 * EXP(22.024 * (1 - 279.24 / ta))
END IF

pa1 = ta / (1 - .3785 * ea / pa)
vo1 = (voo - .4 * voo * (dm / m) + .001 * voo * (tp - Tc))
y0 = SQR(vo1 ^ 2 + wind ^ 2 - 2 * vo1 * wind * COS(a * 3.141592654# / 180))
y0 = y0 * COS(a * 3.141592654# / 180)
z0 = TAN(a * 3.141592654# / 180)
z0 = z0 / (1 - wind / (vo1 * COS(a * 3.141592654# / 180)))
CLS
END SUB


SUB menu (cog, cof, xf, yf, xfu, yfu, t$)
COLOR cog, cof
LOCATE xf - 1, yf: PRINT t$
LOCATE xf, yf: PRINT "É" + STRING$(yfu - yf, 205) + "»";
FOR i = xf + 1 TO xfu
LOCATE i, yf: PRINT "º" + SPACE$(yfu - yf) + "º";
NEXT
    LOCATE xfu + 1, yf: PRINT "È" + STRING$(yfu - yf, 205) + "¼";
END SUB


SUB NPkoef (k, L, r, q, h, y1, z1, v1, w1)
k = h * y1: L = h * z1
r = h * v1: q = h * w1
END SUB
```

```
SUB NPxyzvw (nk, x, x0, y, y0, z, z0, v, v0, w, w0, h, h0, k, L, r, q)
IF nk = 1 THEN
   x = x0: y = y0: z = z0
   v = v0: w = w0: h = h0
   GOTO fund:
END IF

IF nk = 2 OR nk = 3 THEN
   x = x0 + (.5 * h): y = y0 + (.5 * k)
   z = z0 + (.5 * L): v = v0 + (.5 * r)
   w = w0 + (.5 * q)
   GOTO fund:
END IF
IF nk = 4 THEN
   x = x0 + h: y = y0 + k: z = z0 + L
   v = v0 + r: w = w0 + q
END IF
fund:
END SUB


SUB y1z1v1w1 (x, y, z, v, w, y1, z1, v1, w1, koef, pa1, wind, ys, yy, pa, ta1, TE, E, F, D, Pr)
ta1 = (TE / pa1) ^ .5
yy = y * SQR(1 + z ^ 2)
IF yy * ta1 > 256! THEN
   y1 = -1 * koef * (pa / Pr) * ta1 * ((pa1 - .006328 * v) / pa1) ^ 4.4 * (E * yy - F) / yy
ELSE
   y1 = -1 * koef * (pa / Pr) * ta1 * ((pa1 - .006328 * v) / pa1) ^ 4.4 * D * (yy) ^ 2 / yy
END IF
z1 = -9.80665 / y ^ 2
v1 = z
w1 = 1 / y

END SUB
```

APPENDIX F

Algorithm of Improved Euler's Method Applied to Predict the Projectile Trajectory

Consider the **system of differential equations**

$$\begin{cases} \dfrac{dv_x}{dx} = -ch(y) \cdot \dfrac{G_D(v)}{v} \\ \dfrac{dp}{dx} = -\dfrac{g}{v_x^2} \\ \dfrac{dt}{dx} = \dfrac{1}{v_x} \\ \dfrac{dy}{dx} = p \end{cases} \quad (1)$$

where $p = v_y / v_x = \tan\alpha$; $v_x = v \cdot \cos\alpha$, $v_y = v \cdot \sin\alpha$ are the components of velocity. α is the angle the velocity (along the tangent on the trajectory) forms with x-axis, t is the time of flight, $g = 9.80665 m/s^2$, (x, y) are the coordinates of a the projectile moving along the trajectory of flight.

IEM Algorithm

Illustration: 0.338 Lapua GB528 Scenar 19.44 g, cal. 8.59mm bullet

For the ICAO Atmosphere the characteristics G- function of resistance and the ballistics coefficient are respectively:
$$G(v) = 0.141 \cdot v - 30.031, \tag{2}$$

and $c = 3.796$.

For the trajectory of projectile in absence of drag, the ballistic coefficient is $c = 0$.

For shooting in standard conditions at the sea level, we can consider that the density function is one, i.e. $h(y) = 1$.

Initial conditions

$x_0 = 0$, $y_0 = 0$, $t_0 = 0$, departure velocity $v_0 = 830$, departure angle: $\alpha_0 = 1.1471°$.

Derived initial parameters:

$$p_0 = \tan \alpha_0 = \tan(1.1471°) = 0.02002335.$$
$$v_{x0} = v_0 \cos \alpha_0 = 830 \cdot \cos(1.1471) = 829.83366,$$
$$v_{y0} = v_0 \sin \alpha_0 = 830 \cdot \sin(1.1471) = 16.616048$$

Using a "small" step $h = dx$ (in our case $h = 100m$), we can write the iterative formulas associated with system (1).

Euler's Method (Iterative Formulas)

For $n = 0, 1, 2, ..., k-1$, where $k = \dfrac{1500}{h}$ is the number of steps, i.e. subintervals

(in our case $k = \dfrac{1500}{100} = 15$).

The range 1500 meters must be divisible by h in order to end up exactly at x = 1500 meters.

We can write:

$x_{n+1} = x_n + h$, n = 0, 1, ..., 14

$$\begin{cases} v*_{x(n+1)} = v_{x(n)} - c \cdot \dfrac{0.141 \cdot v_n - 30.031}{v_n} \cdot h \\\\ p*_{(n+1)} = p_n - \dfrac{g}{v_{x(n)}^2} \cdot h \\\\ t*_{(n+1)} = t_n + \dfrac{1}{v_{x(n)}} \cdot h \\\\ y*_{(n+1)} = y_{(n)} + p_{(n)} \cdot h \\\\ v*_{(n+1)} = v*_{x(n+1)} \cdot (1 + p*_{(n+1)}^2)^{1/2} \end{cases} \qquad (2)$$

The above formulas are applied in column C of Excel sheet)

$$\begin{cases} v_{x(n+1)} = v_{x(n)} - c \cdot \dfrac{1}{2}(\dfrac{0.141 \cdot v_n - 30.031}{v_n} + \dfrac{0.141 \cdot v*_{(n+1)} - 30.031}{v*_{(n+1)}}) \cdot h \\\\ p_{n+1} = p_n - \dfrac{1}{2}(\dfrac{g}{v_{x(n)}^2} + \dfrac{g}{v*_{x(n+1)}^2}) \cdot h \\\\ t_{(n+1)} = t_n + \dfrac{1}{2}(\dfrac{1}{v_{x(n)}} + \dfrac{1}{v*_{x(n+1)}}) \cdot h \\\\ y_{(n+1)} = y_{(n)} + \dfrac{1}{2}(p_{(n)} + p*_{(n+1)}) \cdot h \\\\ \text{The velocity is:} \\ v_{(n+1)} = v_{x(n+1)} \cdot (1 + p_{(n+1)}^2)^{1/2} \\ \text{Angle of Flight is:} \\ \alpha_{n+1} = \tan^{-1}(p_{n+1}) \end{cases} \qquad (3)$$

$$Drop_{(n+1)} = x_{n+1} \cdot p_0 - y_{(n+1)}$$

Display: x_{n+1}, y_{n+1}, $Drop_{(n+1)}$, v_{n+1}, t_{n+1} **for ranges:** x_{n+1}: **100, 200, ..., 1500 meters.**

The above formulas are applied in column E of excel spreadsheet.

Important note: Formula 4 of system (3) must be applied before formula 2 of the same system.

Improved Euler's Method (Numeric Example)

Substituting $n = 0$ in (2) and after that in (3) we have respectively:

x-coordinate: $x_1 = 0 + 100 = 100$

$$\begin{cases} v*_{x(1)} = 829.83366 - 3.796 \cdot \dfrac{0.141 \cdot (830) - 30.031}{830}(100) = 790.044722 \\ \\ p*_1 = 0.02002335 - 9.80665 \cdot (829.83366)^{-2}(100) = 0.01859925 \\ \\ t*_{(1)} = 0 + 829.83366^{-1} \cdot (100) = 0.12050608 \\ \\ y*_{(1)} = 0 + 0.02002335 \cdot (100) = 2.0023347 \\ \\ v*_1 = (790.044722) \cdot (1 + (0.01859926^2))^{1/2} = 790.1813611 \end{cases} \quad (4)$$

$$\begin{cases} v_{x(1)} = 829.83366 - 3.796 \cdot \dfrac{1}{2}(\dfrac{0.141 \cdot (830) - 30.031}{830} + \dfrac{0.141 \cdot 790.1813611 - 30.031}{790.1813611}) \cdot (100) = \\ = 790.390779 \\ y_1 = y_0 + \dfrac{1}{2} \cdot (0.02002335 + 0.01859926) \cdot 100 = 1.931130 \\ p_1 = 0.02002335 - \dfrac{1}{2}(\dfrac{9.80665}{829.8337^2} + \dfrac{9.80665}{790.390779^2}) \cdot (100) = 0.018525724 \\ t_{(1)} = 0 + \dfrac{1}{2}(\dfrac{1}{829.8337} + \dfrac{1}{790.390779}) \cdot (100) = 0.12354 \\ \text{Ter min } al \text{ velocity and ter min } al \text{ angle are}: \\ v_{(1)} = 790.390779 \cdot (1 + 0.0185257^2)^{1/2} = 790.5264 \\ \alpha_1 = \tan^{-1}(0.018526) = 1.06134020° \end{cases} \quad (5)$$

$$Drop(1) = 100 \cdot (0.02002335) - (1.931130) = 0.0712.$$

The iteration process continues substituting n =1 in (2) and (3) using results in (5), and so on till n=14.

Note: The above math operations (for $n = 0$) can be used to test the programming code.

Improved Euler's method (IEM), C code (IEM.exe)
Prepared by Erio Klimi

The PC program is electronically available to the reader by request

```
// Results.h
#pragma once
#include "IEM.h"
struct Results
{
        double vxStar;
        double vStar;
        double pStar;
        double tStar;
        double yStar;
        double vxN;
        double p;
        double t;
        double Y;
        double vn;
        double drop;
        double A;
};
/**************************************************************/

//IEM.h
#ifndef IEM_H
#define IEM_H
#include<math.h>
#include"Results.h"

//const double C = 3.796;
const double G = 9.80665;
const double PI = 3.14159265;
const double alpha0 = 1.1471 * PI / 180.0;

class IEM
{
public:
        IEM();
        ~IEM();
        void    execute();
        void    output(Results &results);
        void    seth(double _h);
        void    reset();
```

```
            int              getk();
            void     setvx();
            void     setC(double C);
            double   getDistance();
private:
            void     _init_();
private:
            double   h;              int           k;
            double   range;  double  distance;
            double   drop;   double  A;
            double   C;

private:
            double vx;       double vv;
            double p;        double t;

            double vxStar;   double vStar;
            double pStar;    double tStar;
            double yStar;

            double vn;       double vxN;
            double Y;

            int valid;
};
#endif
```

/***/
//IEM.cpp

#include»IEM.h»

IEM::IEM()
{
        init();
}
IEM::~IEM(){}

void IEM::_init_()
{
        h   = 0;   range   = 1500;
        k   = 0;   distance = 0;

```
        drop     = 0;      A = 0;
        C        = 0;

        vx       = 0; vv  = 0;
        p        = 0; t   = 0;
        vxStar   = 0;      vStar = 0;
        pStar    = 0;      tStar = 0;
        yStar    = 0;

        vn       = 0; vxN = 0;
        Y        = 0;

        vx       = 830.0 * cos(alpha0);
        vn       = 830.0;
        p        = tan(alpha0);

        valid = 0;
}

void IEM::execute()
{
        distance += h;

        vxStar = vx - C * (0.141 * vn - 30.031) * h / vn;
        pStar = p - G * h / pow(vx, 2);
        tStar = t + h / vx;
        yStar = Y + p * h;
        vStar = vxStar * pow((1 + pow(pStar, 2)), 0.5);
        vxN = vx - C * 0.5 * ((0.141 * vn - 30.031) / vn + (0.141 * vStar - 30.031) / vStar) * h;

        Y = Y + 0.5 * (p + pStar) * h;
        p = p - 0.5 * (G / pow(vx, 2) + G / pow(vxStar, 2)) * h;
        t = t + 0.5 * (1 / vx + 1 / vxStar) * h;
        vn = vxN * pow((1 + pow(p, 2)), 0.5);
        drop = distance * tan(alpha0) - Y;
        A = 180.0 * atan(p) / PI;
}

void    IEM::seth(double _h) {h = _h; k = (int)(range / h);}
void    IEM::reset() {_init_();}
int     IEM::getk() {return k;}
void    IEM::setvx() {vx = vxN;}
void    IEM::setC(double c) {C = c;}
double  IEM::getDistance() {return distance;}
```

```cpp
void IEM::output(Results &results)
{
        results.drop        = drop;
        results.vxStar      = vxStar;
        results.vStar       = vStar;
        results.pStar       = pStar;
        results.tStar       = tStar;
        results.yStar       = yStar;
        results.vxN                     = vxN;
        results.p           = p;
        results.t           = t;
        results.Y           = Y;
        results.vn                      = vn;
        results.A           = A;
}
/*******************************************************************/
//dll_main.cpp

#include "IEM.h"
#include"Results.h"

#define C_SHARP_CALL __stdcall
#define EXPORT_DLL __declspec(dllexport)

IEM iem;

extern "C" EXPORT_DLL void C_SHARP_CALL IEM_Execution(Results &result)
{
        iem.execute();
        iem.output(result/*out*/);
        iem.setvx();
}
extern "C" EXPORT_DLL void C_SHARP_CALL SetH(double _h) {iem.seth(_h);}

extern "C" EXPORT_DLL void C_SHARP_CALL Reset() {iem.reset();}

extern "C" EXPORT_DLL int C_SHARP_CALL Getk() {return iem.getk();}

extern "C" EXPORT_DLL void C_SHARP_CALL SetC(double c) {iem.setC(c);}

extern "C" EXPORT_DLL double C_SHARP_CALL GetDistance() {return iem.getDistance();}
```

APPENDIX G

System of Differential Equations of Projectile Motion

Improved Euler Method
Applied In Exterior Ballistics

Solution of Differential Equations of Projectile Flight
George Klimi
25 January, 2013

Equation #

$$\frac{dv_x}{dx} = -c \cdot h(y) \cdot \frac{G_D(v)}{v} \quad \text{1}$$

$$\frac{dp}{dx} = -\frac{g}{v_x^2} \quad \text{2}$$

$$\frac{dt}{dx} = \frac{1}{v_x} \quad \text{3}$$

$$\frac{dy}{dx} = p \quad \text{4}$$

Lapua Scenar GB528 19.44 g (300 gr)
Initial Values:

Departure angle = 1.1471 degree= **0.0200206** radian
Step Size, h= **100** m
Form Factor i=1; Ballistic Coefficient, c= **3.796**
Launching Velocity, Vo= **830** m/s
Departure Angle, Alfa(o)= **0.0200206** radian
p0= 0.020023275

Note: Departure angle can be also 0, i.e substitute 0 in F13

Where:
$h(y) = 1$ $\quad p = v_y/v_x = \tan\alpha$

Drag Function of Lapua Bullet:
ICAO Atmosphere
$G(v) = 0.141\ v - 30.031$

v, Projectile Speed; t, time of flight

			Right Side value of 1st equation of System	Right Side value of 1st equation of System

Improved Euler's method

$x(n+1) = x(n)+h \quad Y(n+1) = y(n)+h[f(x(n), y(n)] \quad y(n+1)=y(n)+h\ [\ f(x(n),\ y(n)) + f(x(n+1),\ \dot{Y}(n+1)]/2$

n					
0	100 m (range)				
(Equation #1)	v*x	790.0447	vx	790.4692 m/s	0.3979
(Equation #2)	p*	0.0186	p	0.0185	
(Equation #3)	t*	0.1205	t	0.1235 s	
(Equation #4)	y*	2.0023	y	1.9311 m; y-coordinate	
Projectile Velocity	v*	789.9081	v	790.3336 m/s	
		Projectile	Drop=	0.0712	0.3910
1	200 m (range)				
	v*x	751.2340	vx	751.7460	0.3835
	p*	0.0170	p	0.0169	
Time	t*	0.2500	t	0.2534	
Y-coordinate	y*	3.7837	y	3.7052	
Projectile Velocity	v*	751.1260	v	751.6391	
			Drop=	0.2994	0.3836

2	300 m (range)		713.2820		
	v*x		713.2820	vx	713.7977
	p*		0.0151	p	0.0150
	t*		0.3864	t	0.3900
	y*		5.3924	y	5.3057
	v*		713.2003	v	713.7170
		Drop=			0.3755	0.3754
3	400 m (range)				
	v*x		676.1658	vx	676.6907
	p*		0.0131	p	0.0130
	t*		0.5301	t	0.5340
	y*		6.8097	y	6.7135
	v*		676.1076	v	676.6335
		Drop=	**0.7013**		0.3668	0.3666
4	500 m (range)				
	v*x		639.9577	vx	640.4982
	p*		0.0109	p	0.0107
	t*		0.6817	t	0.6860
	y*		8.0141	y	7.9070
	v*		639.9199	v	640.4613
		Drop=	**1.2958**		0.3572	0.3571
5	600 m (range)				
	v*x		604.7370	vx	605.3000
	p*		0.0083	p	0.0082
	t*		0.8421	t	0.8467
	y*		8.9808	y	8.8612
	v*		604.7159	v	605.2796
		Drop=	**2.1046**		0.3469	0.3467
6	700				
	v*x		570.5899	vx	571.1829
	p*		0.0055	p	0.0054
	t*		1.0119	t	1.0170
	y*		9.6814	y	9.5476
	v*		570.5812	v	571.1747
		Drop=	**3.1527**		0.3357	0.3354
					4.4687

7	800				
	v*x	537.6096	vx	538.2409	
	p*	0.0024	p	0.0022	
	t*	1.1920	t	1.1975	
	y*	10.0833	y	9.9330	0.3232
	v*	537.6081	v	538.2396	
			Drop=	6.0856	0.3234
8	900				
	v*x	505.8957	vx	506.5740	
	p*	-0.0012	p	-0.0015	
	t*	1.3833	t	1.3892	
	y*	10.1488	y	9.9796	0.3099
	v*	505.8954	v	506.5735	
	Range 900 m		Drop=	8.0414 m	0.3102
9	1000				
	v*x	475.5536	vx	476.2882	
	p*	-0.0053	p	-0.0055	
	t*	1.5866	t	1.5931	
	y*	9.8345	y	9.6435	0.2955
	v*	475.5470	v	476.2810	
			Drop=	10.3798	0.2959
10	1100				
	v*x	446.6923	vx	447.4929	
	p*	-0.0099	p	-0.0101	
	t*	1.8030	t	1.8100	
	y*	9.0905	y	8.8744	0.2800
	v*	446.6706	v	447.4699	
			Drop=	13.1512	0.2805
11	1200				
	v*x	419.4224	vx	420.2988	
	p*	-0.0150	p	-0.0154	
	t*	2.0334	t	2.0409	
	y*	7.8595	y	7.6147	0.2634
	v*	419.3749	v	420.2490	
			Drop=	16.4132	0.2640

12	1300			
		393.8516	vx	0.2457
		-0.0209	p*	
		2.2789	t*	
		6.0763	y*	
		393.7654	v*	
				0.2464
			Drop=	20.2316
13	1400			
		370.0806	v*x	0.2271
		-0.0276	p*	
		2.5401	t*	
		3.6666	y*	
		369.9396	v*	
				0.2280
		Range 1400m	Drop=	24.6805
14	1500			
		348.1961	vx	0.2076
		-0.0352	p	
		2.8180	t	
		0.5474	y	
		347.9810	v	
				0.2087
		RANGE 1500m	DROP=	29.8435 meters
	1600	328.2639038		0.1876
		-0.043685566		
		3.113143776		
		-3.373660463		
		327.9511173		
				0.1890

Impact Time   349.3580 m/s
               -0.0357 radian
Y-coordinate   2.8269 s
Impact Velocity   0.1914 m
Impact Angel (degree)`   349.1362 m/s
              -2.041768902

              329.5403
              -0.0442
              3.1223
              -3.7754
              329.2186

Not Reliable data

(1) Refer to http://en.wikipedia.org/wiki/External_ballistics

(2) Drag Function of Lapua Bullet (ICAO Atmosphere):
   $G(v) = 0.141 * v - 30.031$
   Ref. Equation 1.7.33, page 95: Exterior ballistics: The Remarkable Methods, Klimi, G, Xlibris 2014

(3) System of equations of Projectile Motion (1, 2, 3, 4):
   Ref. Equation 2.7.16, page 100: Exterior Ballistics with Applications, 3rd. ed., Klimi, G., Xlibris 2011

Note: Bullet Drop

$$Drop = range \times \tan(\alpha_0) - y$$

where α_0 is the departure angle;
while y is the y-coordinate at range x = 900

Example 1. Drop at the range 900 meters is:

$$Drop_{900} = 900 \cdot \tan(1.1471) - 9.979578117 = 8.0414 m$$

COMPARE THE ABOVE RESULTS WITH WIKIPEDIA (LAPUA) DATA
Departure Angle: 1.1471 degree

Wikipedia Data, Lapua Scenar GB528 19.44 g (300 gr)

Range (m)	0	300	600	900	1,200	1,500
Velocity (m/s)	830	711	604	507	422	349
Time (s)	0	0.3918	0.8507	1.3937	2.0435	2.8276
Bullet Drop (m)	0	0.715	3.203	8.146	16.571	30.035

System of Differential Equations of Projectile Motion

$$\begin{cases} \dfrac{dv_x}{dx} = -c \cdot h(y) \cdot \dfrac{G_D(v)}{v} & \text{Equation \#} \\[4pt] \dfrac{dp}{dx} = -\dfrac{g}{v_x^2} & 1 \\[4pt] \dfrac{dt}{dx} = \dfrac{1}{v_x} & 2 \\[4pt] \dfrac{dy}{dx} = p & 3 \\ & 4 \end{cases}$$

Where:
$h(y) = 1$ $p = v_y/v_x = \tan\alpha$

Drag Function of
Lapua Bullet: $G(v) = 0.141\ v - 30.031$
ICAO Atmosphere

v, Projectile Speed; t, time of flight

Improved Euler Method
Applied in Exterior Ballistics
Solution of Differential Equations of Projectile Flight

George Klimi
30-May-15

Lapua Scenar GB528 19.44 g (300 gr)
Initial Values:

Departure angle = 1.1471 degree=	0.0200206	radian
Step Size, h=	100	m
Form Facor i=1; Ballistic Coefficient, c=	3.796	
Launching Velocity, Vo=	830	m/s
Departure Angle, Alfa(o)=	0.0200206	radian
p0=	0.020023275	

Note: Departure angle can be also 0, i.e substitute 0 in F13

Right Side value of 1st equation of System

Euler's method

$n \quad x(n+1) = x(n)+h \quad \hat{Y}(n+1) = y(n)+h[\ f(x(n), y(n)]$

0.0000	100.0000 m (range)			
(Equation #1)	v*x	790.0447	vx	0.3979
(Equation #2)	p*	0.0186	p	
(Equation #3)	t*	0.1205	t	
(Equation #4)	y*	2.0023	y	
Projectile Velocity	v*	789.9081	v	
		Projectile	Drop=	0.0000
1.0000	200.0000 m (range)			
	v*x	750.9529	vx	0.3909
	p*	0.0170	p	
Time	t*	0.2471	t	
Y-coordinate	y*	3.8622	y	
Projectile Velocity	v*	750.8440	v	

2.0000 | 300.0000 m (range) | | |
v*x	712.6119	vx
p*	0.0153	p
t*	0.2471	t
y*	5.5650	y
v*	712.5286	v
			0.3834

3.0000 | | | Drop= | 0.4419
400.0000 m (range)		
v*x	675.0873	vx
p*	0.0134	p
t*	0.3874	t
y*	7.0940	y
v*	675.0271	v
			0.3752

4.0000 | | | Drop= | 0.9154
500.0000 m (range)		
v*x	638.4516	vx
p*	0.0112	p
t*	0.5355	t
y*	8.4297	y
v*	638.4115	v
			0.3664

5.0000 | | | Drop= | 1.5819
600.0000 m (range)		
v*x	602.7844	vx
p*	0.0088	p
t*	0.6922	t
y*	9.5504	y
v*	602.7611	v
			0.3567

6.0000 | | | Drop= | 2.4636
700.0000		
v*x	568.1734	vx
p*	0.0061	p
t*	0.8581	t
y*	10.4304	y
v*	568.1629	v
			0.3461

Drop= 3.5859

7.0000	800.0000		534.7141	vx	0.3346
		v*x	0.0031	p	
		p*	1.0341	t	
		t*	11.0405	y	
		y*	534.7116	v	
		v*			
				Drop=	4.9781
8.0000	900.0000		502.5100	vx	0.3220
		v*x	-0.0004	p	
		p*	1.2211	t	
		t*	11.3469	y	
		y*	502.5099	v	
		v*			
		Range 900 m		Drop=	6.6741
9.0000	1000.0000		471.6720	vx	0.3084
		v*x	-0.0042	p	
		p*	1.4201	t	
		t*	11.3102	y	
		y*	471.6678	v	
		v*			
				Drop=	
10.0000	1100.0000		442.3175	vx	0.2935
		v*x	-0.0087	p	
		p*	1.6321	t	
		t*	10.8852	y	
		y*	442.3009	v	
		v*			
				Drop=	11.1404
11.0000	1200.0000		414.5677	vx	0.2775
		v*x	-0.0137	p	
		p*	1.8582	t	
		t*	10.0195	y	
		y*	414.5289	v	
		v*			
				Drop=	14.0085

12.0000	1300.0000		388.5446	vx	0.2602
			-0.0194	p*	
			2.0994	t*	
			8.6524	y*	
			388.4717	v*	
13.0000	1400.0000	Drop=	17.3778		0.2418
			364.3662	vx	
			-0.0259	p*	
			2.3568	t*	
			6.7148	y*	
			364.2443	v*	
		Range 1400m Drop=	21.3178		0.2223
14.0000	**1500.0000**		342.1396	vx	
			-0.0333	p*	
			2.6312	t*	
			4.1276	y*	
			341.9505	v*	
		RANGE 1500m DROP=	25.9074		0.2019
Impact Time			321.9535	vx	
Y-coordinate			-0.0416	p*	
Impact Velocity			2.9235	t*	
npact Angel (degree)=	**0.0000**		0.8017	y*	
			321.6748	v*	
	1600.0000				
	Not Reliable data				

(1) Refer to **http://en.wikipedia.org/wiki/External_ballistics**

(2) Drag Function of Lapua Bullet (ICAO Atmosphere):
$K(v) = 0.141 * v - 30.031$
Ref. Equation 1.7.33, page 95: Exterior ballistics: The Remarkable Methods, Klimi, G, Xlibris 2014

(3) System of equations of Projectile Motion (1, 2, 3, 4):
Ref. Equation 2.7.16, page 100: Exterior Ballistics with Applications, 3rd. ed., Klimi, G,, Xlibris 2011

Note: Bullet Drop

$Drop = range \times \tan(\alpha_0) - y$
where α_0 is the departure angle;
while y is the y-coordinate at range x = 900

Example 1. Drop at the range 900 meters is:

$Drop_{900} = 900 \cdot \tan(1.1471) - 9.979578117 = 8.0414 m$

COMPARE THE ABOVE RESULTS WITH WIKIPEDIA (LAPUA) DATA

Wikipedia Data, Lapua Scenar GB528 19.44 g (300 gr)   Departure Angle: 1.1471 degree

Range (m)	0	300	600	900	1,200	1,500
Velocity (m/s)	830	711	604	507	422	349
Time (s)	0	0.3918	0.8507	1.3937	2.0435	2.8276
Bullet Drop (m)	0	0.715	3.203	8.146	16.571	30.035

APPENDIX H

Electronic Copies of PC Programs

Dear Reader

To request (free) electronic copies of the PC Programs that are included in the book, please send a request message to the following e-mail addresses:

iven24@aol.com.

Visit also the website http://gklimi.wix.com/physicsmathapp/home

DISCLAIMER

Neither the author nor Xlibris accepts any responsibility, or liability for any problem (errors, damages, injuries, etc.) that might occur in practice of shooting using the information given in this book, and by employing the PC programs to solve practical ballistics problems.

The user of the ballistics methods and the associated PC programs that are presented in the book, should be cautious, and use the common sense to judge the predicted outcomes and their practical use.

REFERENCES

1. Carlucci, D. E., Jacobson, S. S., Ballistics: Theory and Design of Guns and Ammunition, 2nd edition, CRC Press, 2014.
2. Cranz, C., Becker, K., Exterior Ballistics, London, 1921
3. Cronander, H.A.N. S., G-Dragfunctions.xls"; G-Dragmodels.xls, http://www.cronander.net/, November 10th, 2005.
4. De Mestre, N., The Mathematics of Projectiles in Sport, Cambridge University Press, 1990.
5. Didion, I., Cours Elémentaire De Balistique, Paris, 1852
6. Engineering Design Handbook, Trajectories, Differential Effects, and Data for Projectiles, U.S. Army Materiel Command, 1963 (unclassified)
7. Dmitrievskij, A. A., Exterior Ballistics, Moscow 1972
8. Field Artillery, Volume 6, DND Canada, 1992 – http://www.scribd.com/doc/4934783/BALLISTICS-AND-AMMUNITION, (Web access December 20, 2009)
9. Gubinim, S. G., Gorovim, S. A. Ballistics, Handbook, http://www.ssga.ru/AllMetodMaterial/metod_mat_for_ioot/metodichki/ballistica/index.htm (web access 2008).
10. Hatcher J. S., "Hatcher's Notebook", Stackpole Books, 1962.
11. Hayden, R, Almgren, T., Thomas, K., McDonald W. T., Exterior Ballistics Explained, 5th Edition, Exterior Ballistics.com.
12. Herrmann, E. E., Exterior Ballistics, U.S. Naval Institute, The College Press, 1935.
13. Hurley, J. P., and Garrod, C., Principi Di Fisica, Zanichelli, 1986.

14. Klimi, G., Exterior Ballistics with Applications – Skydiving, Parachute Fall, Flying Fragments, 1st Edition, Xlibris, 2008.
15. Klimi, G., Exterior Ballistics with Applications: Skydiving, Parachute Fall, Flying Fragments, 3rd edition, Xlibris, 2011
16. Klimi, G., Exterior Ballistics of Small Arms, Xlibris, 2009.
17. Klimi, G., Exterior Ballistics: A New Approach, Xlibris, 2010
18. Klimi, G., Exterior Ballistics: The Remarkable Methods, Xlibris 2014
19. Kneubuehl, B. P., What is the maximum length of a spin stabilized projectile, January1987, published by http://www.researchgate.net/publication/253794843
20. Krasnov, N.F, Aerodynamic of Bodies of Revolution, Elsevier Publishing Inc., NY 1970.
21. Krčmář, Jan. PC Program Ballistica 2.2, http://www.balistika.cz/eng/exterior.html, (accessed on 6 November, 2009).
22. Krčmář, Jan, Exterior Ballistics 2.4, http://www.balistika.cz/eng/exterior.html
23. McCoy, R. L., Modern Exterior Ballistics, Schiffer Publishing Ltd., 1999.
24. McCoy, R. L.,Aerodynamic Characteristics of N Caliber .22 Long Rifle Match Ammunition, BRL 1990 (Unclassified)
25. McDonald, W., Inclined Fire, June, 2003, http://www.exteriorballistics.com/ebexplained/5th/50.cfm
26. McShane, E. J., Kelly, J. L., Reno, F., Exterior Ballistics, The University of Denver Press, 1953.
27. Marvin E. Backman, Terminal Ballistics, Research Department, Naval Weapons Center, February 1976 (Approved for Public release).
28. Miller, D. A New Rule for Estimating Rifling Twist, Precision Shooting, March, 43-48 (2005)
29. Mori, E., Balistica teorica e pratica; http://www.earmi.it/balistica, November 2009.
30. Mountain Range Table of 122mm cannon Mod. 1960 – Projectile OF-472, Ministry of Defense of Albania, Tirana 1972.
31. Mucinov, S.S., Shevcenko, N.A., Zadacnik po Osnovami Strelbi is Strelkovogo Oruzie, 1964.
32. Norwood, John M., Comparison of Approximate Methods for Airborne Gunnery Ballistics Calculations, The University of Texas at Austin, TECHNICAL REPORT AFAL-TR-73-179, April 1973 (unclassified)
33. Okunev, B. H., Fundamentals of Ballistics, Vol.1, Book 2, Moscow, 1943.
34. Plostins, P., McCoy, R. Wagoner, B. A., Aeroballistics Performance of the 25mm M910 TPDS-T Range Limited N Training Projectile, BRL Aberdeen Proving Ground, Maryland, 1991 (Unclassified)

35. Range Tables of Cannon 122mm, Mod. 1960, Ministry of Defense of Albania, Tirana, 1967.
36. Rinker, R. A., Understanding Firearm Ballistics, Mulberry House Publishing, 6th Ed, 2005
37. Robinson, G., Robinson, I, The motion of an arbitrarily rotating spherical projectile and its application to ball games, Online at stacks.iop.org/PhysScr/88/018101
38. Roller, D. E., Blum, R., Fisica, Vol. 1, Zanichelli, 1984.
39. Shapiro, J. M., Vneshnaja Balistika, Oborongiz ', 1946
40. Von Wahlde, R., and Metz, D., Sniper Weapon Fire Control Error Budget Analysis, U.S. Army Research Laboratory, Aberdeen Proving Ground, MD, August 1999 (unclassified, web access July 1st, 2012).
41. Weinacht, P., Cooper, G. R., Newill, J. F. Analytical Prediction of Trajectories for High-Velocity Direct-Fire Munitions, ARL, August 2005, (unclassified, web access July, 2013)
42. Whelan, P. M., Hodgson, M. J., Essential Principles of Physics, J. Murray, 1979.
43. Wikipedia, External Ballistics, http://en.wikipedia.org/wiki/External_ballistics (web access, October 24, 2009, December 2012, August 2013)
44. Wikipedia, http://en.wikipedia.org/wiki/QuickLOAD (web accessed, 08/28/2013)
45. Wikipedia, http://www.lapua.com/en/products/sport-shooting/centerfire-rifle/23 (Web access 12/01/2013)
46. Zill, D. G., Cullen, M. R., Differential Equations with Boundary-Value, 5th Ed., Books/Cole, 2001.

END NOTES

1. (http://www.exteriorballistics.com/ebexplained/4th/54.cfm)
2. https://en.wikipedia.org/wiki/External_ballistics, table at: Predictions of several drag resistance modelling and measuring methods
3. (http://www.lapua.com/en/products/sport-shooting/centerfire-rifle/23)
4. http://www.lapua.com/en/products/sport-shooting/centerfire-rifle/23). lapuaproductcatalogue2012usa.pdf
5. Klimi, G. Exterior Ballistics with Applications, 3rd ed, p. 44 - 48, Xlibris 2011.
6. Klimi, G. Exterior Ballistics with Applications, 3rd ed, p. 44 - 48, Xlibris 2011.
7. Klimi, G. Exterior Ballistics with Applications, 3rd edition, p. 46, Xlibris 2011
8. McCoy, Robert L, Modern Exterior Ballistics, Firing Uphill and Downhill, p.47. Schiffer 1999.
   McDonald, William T, Inclined Fire,
   http://www.exteriorballistics.com/ebexplained/article1.html.
   Peters, V. J., at alt. Ballistic Ranging Methods and Systems for Inclined Shooting, US Patent
   Date February 2, 2010. http://www.stoel.com/webfiles/7654029.pdf, (Accessed 06/12/ 2015).
9. Okunev, B, H. Fundamentals of Ballistics, page. 240, Vol 1, Book 2, Moscow 1943.
   Klimi, G, Exterior Ballistics with Small Arms, (pp. 62, 63: Formulas (1.7.1), (1.7.5), (1.7.6), Xlibris 2009
10. Field Artillery (vol.6), Ballistics and Ammunition, DND Canada, 1992
11. External Ballistics (Predictions of several drag resistance modelling and measuring methods),
    https://en.wikipedia.org/wiki/External_ballistics, accessed 06/14/2015
12. Klimi, G. (2010). Exterior Ballistics: A New Approach, p. 234, Xlibris.

13. Klimi, G.(2011). Exterior Ballistics with Applications (3rd ed.), p. 112. Xlibris
14. Klimi, G., Exterior Ballistics: The Remarkable Methods", table 16, p. 112, Xlibris 2014
15. McCoy, Robert, "Modern Exterior Ballistics", page 101 and page 169, Schiffer Publishing, 1999.
16. Formulas (1.3.7) and (1.3.8) are a courtesy of Prof. James Lewis (Marquette University) he developed using Clausius-Clapeyron equation to approximate water vapor pressures for liquid-vapor equilibrium and solid-vapor equilibrium respectively.
17. Klimi, G., Exterior Ballistics with Applications, 3rd Ed. equation (8.4.11), Xlibris 2011.
18. Klimi, G., Exterior Ballistics with Applications, 3rd edition, section 8.3, Xlibris, 2011
19. Klimi, G., Exterior Ballistics: The remarkable methods, section 7.2, Xlibris, 2014
20. The coauthor of *The Exponential Equation of Bullet Trajectory*, is Mr. Felice Nunziata. **Felice Nunziata** is a physicist at ARPAC (Campania's Regional Environmental Protection Agency) in Naples, and forensic analyst at District Court at the Office of Public Attorney - Italy. The paper is published for the first time at EBA, third ed. 2011.
21. Table 2 can be obtained using PC program QuickTarget Unlimited Lapua Edition. (Refer to http://en.wikipedia.org/wiki/External_ballistics)
22. Shapiro, J. M., Exterior Ballistics, p.205, Moscow 50'
23. Jan Krčmář, Ballistica2.2, Powder Temperature and Barrel length, http://www.balistika.cz/eng/exterior.html (accessed on 6 November, 2009)
24. Rinker, R.A. "Understanding Firearm Ballistics", p.162, 6th Edition, Mulberry House Publishing, 2005
25. McCoy, R., Modern Exterior Ballistics, table 8.9 and table 8.10,1st. Ed., Schiffer Publishing Ltd.,1999
26. McCoy, Robert L. Modern Exterior Ballistics, p.165, Schiffer Publishing, 1999
27. Klimi, G. Exterior Ballistics with Applications, Xlibris 2008; 3rd ed. 2011
    Klimi, G. Exterior Ballistics: A New Approach, Xlibris 2010
    Klimi, G, Exterior ballistics: The Remarkable Methods, Xlibris 2014.
28. Quoted from "Ballistics and Field Artillery", Vol. 6: Ballistics and Ammunition, chapter 6, section 1, (Variations and Corrections), DND Canada, 1992.
29. Klimi, G., Exterior Ballistics; A New Approach, Section 1.5, Xlibris 2010
30. http://en.wikipedia.org/wiki/External_ballistics, Accessed 12/20/2015
31. www.Lapua.com, Lapua Special Purpose, English, web access 11/25/2012
32. See Ref. 4.
33. www.Lapua.com, Lapua Special Purpose, English, web access 11/25/2012. McDonald, W.T, Almgren, T. C. The Ballistic Coefficient, 2008, http://www.exteriorballistics.com/ebexplained/articles/the_ballistic_coefficient.pdf
34. Michael and Amy Courtney, The Truth About Ballistic Coefficients, http://arxiv.org/ftp/arxiv/papers/0705/0705.0389.pdf (accessed 12/26/2015)
35. Shapiro, p. 225, Mc Coy, p. 192
36. Lapua Special Purpose 2012 English.pdf., Issued by Lapua

37 McCoy, Shapiro 1946
38 Shapiro, J. M., Exterior Ballistics, p. 236, Moscow 1946
39 A Study on the Improved Euler's Method in Exterior Ballistics
40 McCoy, Robert., Modern Exterior Ballistics, p.166, Schiffer Publishing 1999.
41 The C/C++ program IEM.exe and the respective code is prepared by my son, Erio Klimi who is a software developer.
42 Klimi, G., Exterior Ballistics: A New Approach, p. 234, Xlibris 2010.
43 Klimi, G. Exterior Ballistics with Applications (3rd ed.), p. 112. Xlibris 2011
44 External Ballistics, [Online]. http://en.wikipedia.org/wiki/External_ballistics, Predictions of several drag resistance modelling and measuring methods (accesed May 29, 2015)
45 Klimi, G. "Exterior Ballistics: The Remarkable Methods, (page 192-200), Xlibris 2014.

www.ingramcontent.com/pod-product-compliance
Lightning Source LLC
Chambersburg PA
CBHW020729180526
45163CB00001B/167